ECHINODERM STUDIES

VOLUME 1

ECHINODERM STUDIES

edited by

MICHEL JANGOUX
Université Libre de Bruxelles, Belgium

JOHN M.LAWRENCE
University of South Florida, Tampa, USA

VOLUME 1

CRC Press
Taylor & Francis Group
Boca Raton London New York

CRC Press is an imprint of the
Taylor & Francis Group, an **informa** business
A BALKEMA BOOK

Published by:
CRC Press/Balkema
P.O. Box 447, 2300 AK Leiden, The Netherlands
e-mail: Pub.NL@taylorandfrancis.com
www.crcpress.com – www.taylorandfrancis.com

CRC Press/Balkema is an imprint of the Taylor & Francis Group, an informa business
No claim to original U.S. Government works

ISBN 13: 978-90-6191-290-3 (hbk)

CONTENTS

PREFACE

Echinoderms have long been studied by scientists, and interest in all aspects of echinoderm biology is continually increasing. These studies make contributions to the fields of ecology, evolution, paleontology, developmental biology, molecular biology, reproduction, behavior and physiology.

The steady advance in the study of echinoderms has led us to believe that the time is appropriate to initiate a series of reviews, scheduled to appear biennially, on echinoderm biology. A guiding principle of the series is to cover all aspects of echinoderm biology, believing as we do that such an approach will promote a better comprehension of this remarkable group of animals.

The editors

PATTERNS AND PROBLEMS IN ECHINODERM EVOLUTION

JAMES SPRINKLE
Department of Geological Sciences, University of Texas, Austin, USA

CONTENTS

1 INTRODUCTION

Echinoderms have a long and complex Phanerozoic fossil record (Fig.1) extending back at least to the Early Cambrian (about 570-600 million years ago) and perhaps even to the latest Precambrian (between 600-700 million years ago). During this long interval of time, many echinoderm classes have appeared in the fossil record, evolved new morphologies and ways of life, and then either continued to successfully diversify or became extinct. Echinoderms today are abundant and diverse in many marine environments, but include only five classes (crinoids, asteroids, ophiuroids, echinoids, and holothuroids). In contrast, at least 15 other classes of echinoderms are extinct and known only from

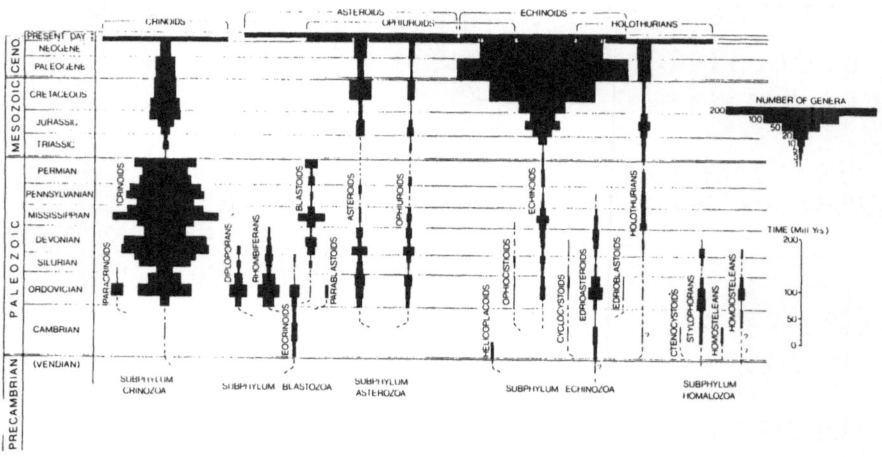

Figure 1. The fossil record of echinoderms showing generic diversity of the 20 classes plotted through geologic time (see scales at right) (from Sprinkle 1980a).

the fossil record. All living echinoderm classes are at or nearly at their maximum generic diversity today (Fig.1). However, this is probably not the true pattern but a reflection of our incomplete knowledge of the fossil record. Many living echinoderms are poorly skeletized (such as holothuroids), disarticulate rapidly after death (such as asteroids, ophiuroids, and crinoids), or live in environments rarely found in the fossil record (such as the deep sea). Because of these factors, present day echinoderm faunas are much better known (perhaps an order of magnitude better known) than faunas from any period in the Phanerozoic fossil record. This suggests that echinoderm faunas at several times in the past may have been as abundant and diverse as present-day echinoderm fauna. We know only a small percentage of echinoderm fossil diversity even after nearly 200 years of taxonomic work on the fossil record; this is the sample with which we must work.

Even with this incomplete sample, several major patterns in the echinoderm fossil record are evident and several major problems in explaining these patterns or other features of echinoderm evolution are known. I have chosen to discuss four evolutionary patterns and five unresolved problems.

2 EVOLUTIONARY PATTERNS IN ECHINODERMS

2.1 *Initial radiation and appearance of new classes*

One of the most striking features shown by the echinoderm fossil record (Fig.1) is that all 20 classes that are usually recognized (Sprinkle 1980a) appeared in an initial explosive radiation between the earliest Cambrian and the Middle Ordovician. Although this includes an interval of about 140 million years, it is only about 26% of geologic time in the Phanerozoic record. Most of the major ways of life evolved by echinoderms also appeared during this time period; only deep-burrowing detritus feeding appeared later, probably in the Late Paleozoic or Early Mesozoic (Sprinkle 1980a). Three classes representing two subphyla (Echinozoa and Blastozoa) appeared in the Early Cambrian, about 570 million years ago, almost at the beginning of the metazoan fossil record (Fig.1). Two classes continued, but the third (helicoplacoids) became extinct soon after it appeared, even though it then represented the most diverse group of Early Cambrian echinoderms with three genera and six described species (Durham 1967a). At least four and perhaps as many as six additional echinoderm classes appeared in the Middle Cambrian, adding two more subphyla (Homalozoa and Crinozoa) to the echinoderm record. Two of these classes became extinct within the Middle Cambrian, but the others continued. Only one new class (homoiosteleans) appeared in the Late Cambrian (Fig.1) but there may be earlier records of this group (Derstler 1975, and unpublished). Thus, the Cambrian record includes a total of 8-10 echinoderm classes and four of the five subphyla.

Five new classes and the fifth echinoderm subphylum (Asterozoa) appeared near the base of the Ordovician Period, about 470-480 million years ago. One other class (crinoids) reappeared with much greater abundance after a single isolated occurrence in the Middle Cambrian. All of these classes were fairly successful groups that continued on at least until the Devonian. The last five echinoderm classes appeared in the Middle Ordovician about 440-450 million years ago. These included both early members of large successful groups (such as blastoids and echinoids) and several small and short-lived classes. Two classes also reappeared from earlier occurrences in the Cambrian. The Middle Ordovician marks the high point in terms of number of echinoderm classes existing at any one time (17 classes); as many as 13 of these may occur together in a single stratigraphic unit, such as in the Bromide Formation of Oklahoma (Sprinkle 1982). No new echinoderm classes appeared in the fossil record after the Middle Ordovician (Fig.1), and the rest of the Paleozoic shows a gradual but continuous decrease in the number of echinoderm classes present (Sprinkle 1980a).

This pattern of an initial radiation of numerous higher taxa (such as classes) followed by a gradual drop in class diversity seems to be a common feature of most metazoan phyla (Valentine 1969). Paul (1979) has referred to this as the colonization-radiation/competition-retrenchment model. Lower taxa (such as species, genera, and families) in these phyla show a later peak in diversity and in most cases are at their highest levels in the present-day fauna (Fig.1). Echinoderms have a larger number of classes early in the fossil record and a larger dropoff in class diversity later in the record than most other phyla, but are otherwise similar in their general pattern.

2.2 *Later 'weeding-out' of small and inefficient classes*

The gradual decrease in number of classes after the Middle Ordovician peak until the end of the Paleozoic is another characteristic pattern of the echinoderm fossil record (Fig.1). Class diversity dropped from 17 in the Middle Ordovician, to 13 in the Late Silurian, to 9 in the Late Devonian, to 7 in the Mississippian, and to 6 in the Pennsylvanian and Permian. All five classes that survived the extinction at the Permo-Triassic boundary (about 225 million years ago) are still living today (Fig.1). This decrease in class diversity was not matched by a decrease in the number of genera or species (Sprinkle 1980a) which showed three widely spaced peaks in diversity during the Paleozoic, and remained at fairly high levels until almost the end of the Permian. Why did the number of echinoderm classes eventually decrease almost to the level found in the Early Cambrian?

The answer probably lies in the nature of the ecosystem in which these early and later echinoderms were living and the relative efficiencies of the echinoderms themselves. The earliest echinoderms probably evolved in an ecosystem that was almost devoid of metazoans, had high productivity, and many favorable living areas (Sprinkle 1976). These environmental conditions favored the rapid divergence of widely different types of echinoderms that occupied the available open niches on the margins of widely dispersed continents (Ziegler *et al.* 1979). Because there was little competition from other metazoans, most of these early echinoderms were probably generalists in their way of life. Separate continents apparently had their own endemic faunas, because the distribution of Cambrian echinoderm genera seems very localized. These factors indicate that most Cambrian and Ordovician echinoderms were fairly generalized and inefficient compared to later forms. Extinction of groups was probably common, and because most classes had low diversity and a limited distribution, even classes commonly became extinct.

As diversity of echinoderms and other metazoans increased, competition and predation increased, promoting greater specialization, greater efficiency at using the available resources, and greater emphasis on protection and movement. Classes surviving more than a single geological period show conspicuous improvements in morphology (see section 2.3) and movement into specialized less-crowded niches. Because of these factors, many classes of echinoderms that were slowly-evolving, inefficient or having much competition in their way of life, of low diversity, or limited either geographically or ecologically became extinct one after another in the more crowded and specialized ecosystems of the Middle and Late Paleozoic. This 'weeding out' process mostly affected classes of blastozoan, echinozoan, and homalozoan echinoderms (Fig.1), leaving two of these subphyla extinct by the end of the Paleozoic. Groups becoming extinct in the Middle and Late Paleozoic were either not replaced or were replaced by newly-evolved taxa classified as new families and orders, not new classes (Valen-

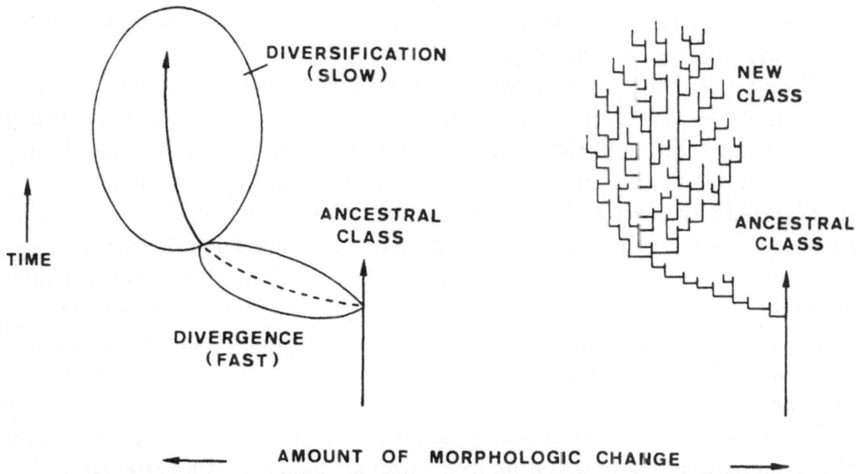

Figure 2. Inferred rates of morphological change in the origin and later evolution of a typical echinoderm class. Left: general pattern showing terms applied to this type of adaptive radiation. Right: same showing amounts of diversity present in each stage.

tine 1969). Thus, the number of echinoderm classes gradually decreased to only five survivors by the beginning of the Mesozoic era.

2.3 *Gradual improvements in morphology in surviving classes*

Most successful echinoderm classes in the fossil record show a characteristic pattern during their history. They appear suddenly with a new morphology different from any possible ancestor; this probably implies a short period of rapid evolutionary change (Fig.2) for which we usually have no preserved fossil record. Successful classes that show some diversity and last for a geologic period or longer usually have somewhat slower rates of evolution and only slight modifications of the basic design for the class. In other words, the origin of a new echinoderm class involved the development of a new and quite different design from the nearest preserved ancestor (probably a rapid event), followed by considerable modification and 'fine tuning' of this basic design to produce a larger number of better adapted and more specialized members, usually a slower and less drastic set of morphologic changes. Some of these later modifications could be considered improvements in morphology that resulted in a more efficient organism. For example, two major improvements made by Paleozoic echinoids include standardization of test designs at two rows each of ambulacral and interambulacral plates plus the development of one large protective spine on each plate (Kier 1965). Permian cidaroids with this advanced design survived the Permo-Triassic extinction (Fig.1) at very low diversity, even though many larger and more successful Permian groups became extinct. This advanced plating design thus became the basis for all later echinoids, both regular and irregular.

Often a morphologic feature was improved and standardized at the time of appearance of a new class. For example, blastoids have very characteristic thecal plating which is different from nearly all other blastozoan or crinozoan echinoderms. Early blastoids appeared in the record with this characteristic plating (17 plates in four circlets plus several small anal deltoid plates) already present. All later blastoids are based on this same plating formula. Even though the thecal shape was greatly modified, including the development of asymmetrical recumbent forms, major thecal plates were rarely lost and no new ones were gained; only the plate shapes and proportions were changed. Improvements in this case involved minor modifications of a standardized thecal design which characterizes the class.

Development of new advanced designs in echinoderms and the modifications of previously existing designs produced better adapted and more efficient echinoderms during the Paleozoic. Older classes with archaic or inefficient designs that could not be sufficiently improved could not compete with these more advanced groups and gradually became extinct. Thus, the gradual improvement in morphology during the Paleozoic is in part responsible for the persistent drop in the number of echinoderm classes.

2.4 *Convergence toward several optimal designs*

Several morphologic designs and ways of life apparently represent adaptative highs that have been colonized by several related or unrelated groups of echinoderms at different times in the past. Some of these forms show such close convergence in morphology that they are difficult to separate and at times in the past have caused problems in classification (Regnéll 1960). Usually only one or two members of a class will converge on a design used by a larger class as its basic way of life, but sometimes two entire classes will show convergence on a single design. The 'blastoid design' representing medium- to high-level, stemmed, suspension feeders having a compact bud-shaped theca, few circlets of plates, well-developed pentameral symmetry, and five ambulacra with feeding appendages is perhaps the best known example (Kesling 1967, Sprinkle 1980b). In addition to many blastoids, members of at least six other classes (four blastozoans, one crinozoan, and one echinozoan), have converged on this design, primarily in the Middle Ordovician, when blastoids first appeared in the fossil record (Broadhead 1980).

A second popular design might be termed the 'carpoid or flat-fish design'; most of these groups were mobile, epifaunal, or barely infaunal, detrital or suspension feeders. This design is characterized by a flattened theca often with a heavier frame of larger marginal plates and either one or two smaller-plated central areas, a larger appendage attached to one end of the theca for movement, sometimes one or two smaller appendages at the other end for feeding, and usually some degree of bilateral symmetry through the theca. Most homalozoans ('carpoids'), including stylophorans, homoiosteleans, homosteleans, and perhaps ctenocystoids, show this design, along with a few blastozoans such as pleurocystitid rhombiferans (Paul 1967a) and perhaps some advanced eocrinoids such as *Lingulocystis* (Ubaghs 1960). These forms probably wriggled or crawled slowly across the sea floor either collecting detrital material from the top layer of sediment or capturing suspended material from sea water just above the bottom. The ability to uncover themselves from sediment dumped by storms or turbidity currents probably was a major requirement for this way of life. All of these groups are Early to Middle Paleozoic in age, and few if any echinoderms have used this design since the last homalozoan groups became extinct in the Middle Devonian.

A third favorable design might be termed the 'turret design'. These forms are low-level, attached, suspension feeders characterized by a heavily calcified, squat, cylindrical theca attached to the substrate and capped by a domed summit with straight ambulacra or short erectile arms for feeding. This design has been used by Early Paleozoic cyathocystids and perhaps a few other isorophid edrioasteroids (Bell 1976, Sprinkle & Bell 1978), by a few Late Paleozoic flexible crinoids, and by a few Late Jurassic to Recent articulate crinoids. Protection and tight attachment to objects on the sea floor was probably the main reason for the morphologic features found in these groups. The cyathocystid edrioasteroids with this design appear to have been paedomorphic opportunists that colonized very shallow water, wave-, tide-, or storm-swept, nearshore areas (Sprinkle & Bell 1978).

3 UNRESOLVED PROBLEMS IN ECHINODERM EVOLUTION

3.1 *How did suddenly-appearing new classes originate?*

Most new echinoderm classes originated suddenly in the fossil record, representing a major jump in morphology from any possible ancestral group (Fig.2). Because all of these new classes originated early in the Phanerozoic, one might argue that the jumps in morphology are actually gaps in a relatively poor fossil record for that time, with evolution occurring at relatively normal rates and connecting lineages going back hundreds of millions of years into the Precambrian (see e.g. Durham 1967b). An alternate view takes the fossil record more as it appears, implying that the jumps in morphology are real and that evolution took place very rapidly (probably in local regions) during these transitions. This latter viewpoint agrees more closely with the accepted model of an adaptive radiation (Simpson 1953, Raup & Stanley 1978) and shows many similarities with the punctuational model of how speciation in the fossil record takes place (Eldredge & Gould 1972, Gould & Eldredge 1977). In terms of mechanisms for sudden appearance, most Early and some Middle Cambrian classes probably appeared suddenly in the record when they acquired skeletons, which greatly increased their chance of fossilization. Other classes probably originated by such rapid evolutionary mechanisms as paedomorphosis (Gould 1977), macro-mutations, or changes in regulatory genes (Sprinkle 1980b) that affected a local population in some part of the world. If a change of this type were favorable, it would spread rapidly by migration producing the sudden appearance of a new group locally when the first representatives reached this area. These rapid evolutionary changes may have taken only one to a few generations or several hundred generations, depending on the mechanism used, but they were probably much faster and larger than more normal evolutionary changes seen in later representatives (Fig.2). Because niches were more open and competition between groups less intense in the Early Paleozoic, more of these rapid evolutionary jumps are found in this early part of the fossil record (Cambrian and Ordovician) than later.

3.2 *Are there really 20 classes of echinoderms?*

The division of echinoderms used here has a total of 20 classes (Sprinkle 1980a, and Fig.1). Besides the five large classes of living echinoderms, this classification has a number of fairly successful Paleozoic groups that are now extinct (such as blastoids, edrioasteroids, and stylophorans) and an even larger group of small extinct classes with low diversity, a limited geographic distribution, and a short time range (such as parablastoids, helicoplacoids, and ctenocystoids). The recognition of a large number of echinoderm classes is a fairly recent development. Between 1960 and 1969, at least ten new classes (about one per year) were proposed or raised from lower category ranks; eight of these are still considered valid. Some authors have objected to such a large number of echinoderm classes, considering them oversplit (Beerbower 1968, Breimer & Ubaghs 1974). These objections mostly involve the numerous, small, Early Paleozoic classes with unusual morphology (Fig.1). How reasonable are these unusual small groups as classes?

Most authors consider several factors important in deciding whether a group warrants ranking as a separate class. The most obvious factor is morphologic distinctiveness compared to other related groups. Is the morphology different enough so that the group being considered cannot reasonably be included with its nearest relatives? Another important factor is some measure of 'success' of the group being considered. How much diversity is present at any one time or throughout the group's history? How long does the group last in the fossil record? Finally, and least important here, has the group survived to the Recent where it can be studied in more detail for information on its soft parts, biochemistry, genetics, larval stages, ecology, and behavior that we usually cannot obtain from the fossil record?

Echinoderm classes differ considerably in the degree they satisfy these factors. Crinoids and echinoids are distinctive groups without close relatives; both were very large and dominant groups at different times in the past (Fig.1); they have very long fossil records extending from the Recent back to the Cambrian and Ordovician, respectively; and both are living today and are fairly successful. The other three living echinoderm classes also show many of these features. Some extinct classes, such as blastoids, are also fairly distinctive (although probably less so than crinoids and echinoids), are moderately diverse, have a fairly long fossil record in the Paleozoic, but are extinct today and do not appear to have close living relatives. Even with these limitations, most authors would probably consider groups with these features as reasonable classes. Another seven classes (such as blastoids, edrioasteroids, and stylophorans) show these features, bringing the total number of large and medium-sized echinoderm classes to 12 or 60% of the total. The remaining eight small and short-lived groups (such as parablastoids, helicoplacoids, and ctenocystoids) (40% of the total) are more of a problem to classify. Each usually represents a moderately to very distinctive group without close relatives, but they have very low diversity (some as small as one genus and species), a fairly short stratigraphic range (and geographic distribution), and are now extinct without close living relatives. Do these small groups represent separate classes or should they be combined with other Paleozoic groups? Mayr (1969) argued that higher-level taxa in an optimal classification should be relatively similar in size. Based on this criterion, these small extinct groups of echinoderms probably would not qualify as separate classes. However, grouping these small classes with other medium-sized ones produces heterogeneous associations which do not appear to be natural. Because of this problem, I believe morphological uniqueness should probably be the dominant factor over success in determining what groups early in the record really are separate classes. Indeed, some medium-sized heterogeneous groups in the present classification, such as eocrinoids, may eventually be split up into more homogeneous groups, each of which might qualify as a separate class. If this were done, eocrinoids would probably become four to five separate classes, a solution that still might not be completely satisfactory. In solving one problem (a heterogeneous early group), we may be producing another (a greatly increased number of small extinct classes).

These small but distinct groups found in the Early Paleozoic probably represent unsuccessful experimental groups that show a considerable amount of divergence from their ancestral form but little diversification. However, even without the success, I believe that most of these small groups should continue to be ranked as separate classes. Some may eventually be consolidated with larger classes, but this will probably be equalled by the splitting of larger heterogeneous early groups into several new classes.

3.3 *Did all fossil echinoderms have a water vascular system with tube feet?*

All living echinoderms have a fairly well-developed water vascular system with characteristic tube feet. Many fossil echinoderm groups retain evidence that they also had similar structures. Usually this consists of pores where the tube feet extended through the ambulacral plates, such as in early echinoids (Kier 1965), helicoplacoids (Durham 1967a), and edrioasteroids (Bell & Sprinkle 1978), or sockets where the tube feet were housed externally, such as in stylophorans (Ubaghs 1968) and early asteroids. Tube feet or radial canals are actually preserved in a few early echinoids (Durham 1966b, Paul 1967b) and in the earliest known crinoid (Sprinkle 1973a). Some authors have inferred from this that all fossil echinoderms had a well developed water vascular system with tube feet similar to those of living echinoderms (Hyman 1955, Nichols 1969, Paul 1979).

I have argued that most or all blastozoan (brachiole-bearing) echinoderms and perhaps members of a few other extinct classes had a greatly reduced water vascular system which lacked external tube feet (Sprinkle 1969, 1973a, b, Sprinkle & Robison 1978). Other authors have subsequently proposed that some or all members of various groups lacked these structures also; these include paracrinoids (Parsley & Mintz 1975), isorophid edrioasteroids (Bell 1976, 1977), and many groups of early echinoderms (Haugh & Bell 1980).

Sprinkle (1973a) cited four arguments which indicate that ambulacral tube feet were probably absent in brachiole-bearing echinoderms. The first involves the lack of openings from the thecal interior out to the ambulacral exterior. In most echinoderms, the ring canal of the water vascular system is located internally around the esophagus; either the radial canals themselves or the tube feet which branch off them extend out to the exterior through an opening or set of pores through the ambulacral plates. Almost no brachiole-bearing echinoderm has openings of this type through the ambulacra (Sprinkle 1973a). Breimer & Macurda (1972) postulated that in blastoids the ring canal itself was external to the skeleton around the margin of the mouth, and thus no openings through the skeleton were needed. However, this arrangement is not known in any living echinoderm with a tightly-sutured globular skeleton. This arrangement might be possible in blastoids, where no hydropore leading to the interior is present, but not in related eocrinoids, rhombiferans, and diploporans where a hydropore (implying an internal ring canal) is usually present. The arrangement of the ring canal and tegmen in living crinoids is not well known, but some Ordovician inadunate crinoids have a hydropore on the summit (Sprinkle 1973b, 1982), and some Mississippian camerate crinoids have traces of a specialized internal circumoral ring preserved on the tegmen interior (see Haugh 1973). These arrangements are quite different from that postulated by Breimer & Macurda (1972) for blastoids, and make their proposal of an external ring canal rather unlikely. Therefore, the lack of openings from the coelomic interior out to the ambulacral tracts in blastozoans probably implies that the water vascular system and its tube feet were absent from these areas.

The second argument involves the absence of traces of a radial canal or a groove system in the food grooves of blastozoan brachioles (Sprinkle 1973a). Breimer & Macurda (1972) described an extensive groove- or ductlike system in the ambulacra of blastoids but could not show that this system continues out to the brachioles. No structures resembling this ambulacral groove system have been observed in the brachioles of any blastozoan echinoderm. The brachioles of some species of the Middle Cambrian eocrinoid *Gogia* have a canal on the suture between the brachiolar plates (nervous?) and two small lateral canals in the food groove (perihaemal?) (Sprinkle 1973a), but these are not at all similar to the extensive ambulacral groove system of blastoids. Relatively few brachioles showing the brachiolar food grooves have been examined in blastozoans, but the absence of an ambulacral groove system in those that have been may indicate that a water vascular system with tube feet was lacking in these appendages.

The third argument involves the problem of fitting radial water vessels and tube feet in a brachiolar food groove when the brachiolar cover plates were closed (Sprinkle 1973a). Most blastozoan echinoderms have a small food groove ('brachiolar tract') with cover plates that are nearly flat or only slightly domed when closed (Sprinkle 1973a). The width and height of the brachiolar food groove is only about 0.1 mm in many of these blastozoans, which is smaller than the dimensions of food grooves in the pinnules of most living crinoids. It may not have been physically possible to pack a radial water vessel and tube feet into the brachiolar food grooves of many blastozoans.

A further problem involves the arrangement of ambulacral cover plates protecting the ambulacral food grooves in blastoids. Most blastoid genera have a single biserial set of ambulacral cover plates. Breimer & Macurda (1972) argued that cover plates of this type could be opened (either by muscles or by hydraulic tube feet) to expose the food grooves. However, some spiraculate blastoids, such as the specimen of *Pentremites elongatus* figured by Macurda (1975, Pl.5, Fig.1) have a multiple series of ambulacral cover plates arranged in an irregular pattern over the main food grooves. Blastoids like this probably could not open these irregularly-arranged cover plates; however, an ambulacral groove system is as well developed here as in other blastoid genera. The question arises why tube feet would be present in the ambulacral tracts of a genus such as *Pentremites* if they could not be exposed to the external sea water for feeding and respiration.

The fourth argument involves the repeated development of accessory thecal respiratory structures in blastozoan echinoderms in contrast to their usual absence in groups such as crinoids which had tube feet on their arms (Sprinkle 1973a). If blastozoans were really similar to crinoids and had the ambulacral and brachiolar food grooves possessing thousands of soft hydraulically-operated tube feet, they should have had considerable respiratory exchange surface and probably would not have needed accessory thecal respiratory structures. Yet, over 90 % of all blastozoans have extensively developed pore- or foldlike thecal respiratory structures, unlike most crinoids (probably about 99 %) which lack them. And, significantly, the only crinoids to have accessory respiratory structures (in addition to pores through the anal tube) are early (mostly Ordovician) forms with a large globular or conical calyx and short, simply branched or unbranched, non-pinnulate arms (*Cleiocrinus, Porocrinus, Carabocrinus, Acolocrinus*). These are the types of crinoids that might have needed accessory respiratory structures to supplement the reduced number of tube feet. This contrast in the development of thecal respiratory structures implies a major difference in the water vascular system which normally serves this respiratory function. I would conclude that this 'major difference' was the absence of tube feet in blastozoan food-gathering brachioles (Sprinkle 1973a).

3.4 *Are stylophorans really echinoderms or chordates?*

There is a dispute about the morphology and affinities of an Early and Middle Paleozoic group of echinoderms known as stylophoran 'carpoids'. Most authors have considered these forms as a group of aberrant, asymmetrical, bottom-living echinoderms (Jaekel 1900, Gill & Caster 1960, Ubaghs 1968, 1975), but Jefferies (1967, 1968, 1979, Jefferies & Prokop 1972) has concluded that they are primitive chordates (Subphylum Calcichordata) with echinoderm affinities, representing an intermediate group between these two phyla. Jefferies' arguments are based in large part on interpretations of ridges, grooves, and pores on the inside of stylophoran thecae and reconstruction of the soft parts housed or attached to these structures in terms of structures found in other living chordates. This comparison and reassignment has received little support from echinoderm paleontologists (Nichols 1969, Ubaghs 1975, Sprinkle 1980a; but see Paul 1971) or from vertebrate paleontologists (Denison 1971; but see Eaton 1970). In particular, Ubaghs' (1975) critique of Jefferies' proposal listed four major problems that, in my opinion, make this proposal highly unlikely: 1) all stylophorans have a typical echinoderm-type skeleton of single-crystal, microporous, calcite plates, in great contrast to the much different skeleton developed by chordates; 2) the details of the skeletal plating (elongate marginal plates, tiny-plated central areas, moveable spines, sutural pores, an anal pyramid, and the articulated arm-like appendage) closely resemble other homalozoan echinoderms but not chordates; 3) the 'aulacophore cover plates' are sometimes preserved partly open or wide open but still articulated in some specimens whereas Jefferies argues that they were permanently closed; and 4) Jefferies must invert these calcichordates in mid-evolution (between cornutes and mitrates) for his homologies with other chordates to be valid (see especially Fig. 5 of Jefferies & Prokop 1972). However, comparison of thecal and aulacophore features between cornutes and mitrates indicates that this is unlikely to have occurred.

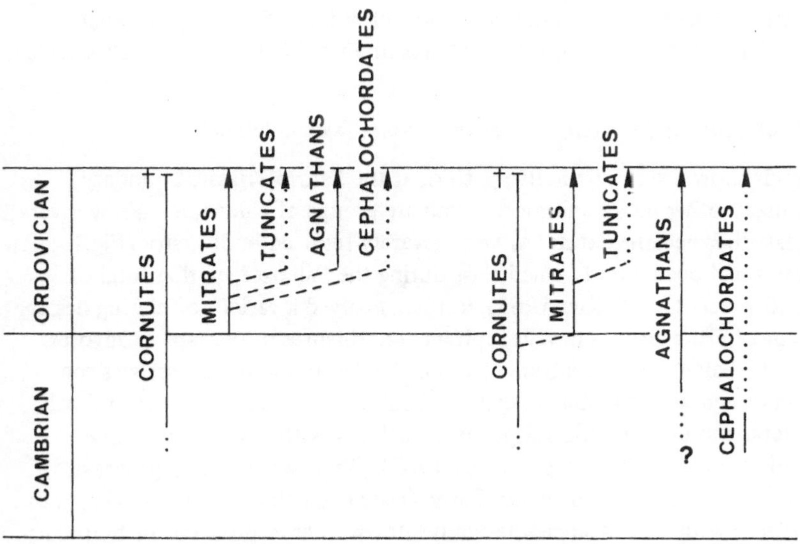

Figure 3. Time of first appearance of the group involved in Jefferies' proposed calcichordate (= carpoid) to chordate phylogeny based on data available in 1967 (left) and in 1981 (right).

Another problem is the time of first appearance of the groups involved. Jefferies (1967) proposed that cornute calcichordates (appearing in the Middle Cambrian) were ancestral to mitrate calcichordates (appearing in the Early Ordovician) which were ancestral to cephalochordates (*Amphioxus*) and to tunicates, both then unknown in the fossil record, and to true vertebrates such as agnathans (appearing in the Early or Middle Ordovician) (Fig.3). No appreciably earlier cornutes or mitrates have been described since 1967, but several reports of primitive vertebrate remains in the Early Ordovician and Late Cambrian (and perhaps even earlier) have appeared (for example see Repetski 1978) although some of these have been questioned (Thompson & Plotnick 1981). Conway Morris has reinterpreted the enigmatic genus *Pikaia* from the Middle Cambrian Burgess Shale of western Canada as a cephalochordate similar to the living *Amphioxus* (see Conway Morris & Whittington 1979). These reports of early chordates make Jefferies' proposed phylogeny (cornutes to mitrates to cephalochordates, tunicates, and agnathans) much less probable (Fig.3).

These problems make Jefferies' morphologic comparisons and taxonomic assignment of stylophorans unlikely. I view stylophorans as an unusual group of mobile, benthic, detritus or suspension feeders (Sprinkle 1976, Parsley & Caster 1975) that have converged on a 'flat-fish' design used by many later bottom-living fish groups. This convergence probably explains many of the similarities between stylophorans and some chordates. Stylophorans appear to be true echinoderms and not intermediates or ancestral to any early chordates.

3.5 *What happened to echinoids at the Permo-Triassic boundary?*

Echinoids show a very distinctive pattern at the Permo-Triassic boundary, where many other larger and more dominate groups of Paleozoic marine invertebrates either became extinct or were severely reduced in diversity (Fig.1, left). Echinoids had been a fairly small class during the Paleozoic with a total of about 40 genera (Fig.1; Kier 1965), but had evolved a variety of plating designs to live most efficiently as mobile, epifaunal herbivores or perhaps carnivores. Late in the Paleozoic, one advanced design began to dominate; this was the cidaroid design with a globular, slightly flexible test having two rows of both ambulacral and interambulacral plates in each ray with a long stout spine attached to many of these plates (Kier 1965). Only two echinoid genera are known from the Late Permian and Early Triassic. Of these, only a single genus (*Miocidaris*) with this advanced design is known to have survived the boundary, while many echinoid competitors and perhaps predators became extinct. Once through the extinction boundary, these surviving cidaroid echinoids had a distinct advantage in these now partly-open environments. By the Late Triassic, several different types of regular echinoids with new morphologic features (gills, compound ambulacral plates, keeled teeth) had evolved from the surviving cidaroids (Kier 1973, 1977b). The first irregular echinoids appeared rather suddenly in the Early Jurassic along with several additional groups of regular echinoids. These earliest irregular echinoids apparently evolved from regular echinoid ancestors in the Late Triassic that had adopted a new way of life by becoming more mobile and plowing through and ingesting the top layer of organic-rich sediment. This change in feeding and way of life was accompanied by major changes in morphology: the anus moved out of the apical system and down the test posteriorly, the test became elongated and more streamlined, the spines became much smaller but more numerous, the peristome and lantern became smaller and began to move anteriorly, and the tube feet became modified for different functions such as respiration and collecting sediment (Kier 1974). These irregular echinoids with their burrowing way of life and new morphology became the dominant echinoids in the Late Mesozoic and Cenozoic.

This record raises several interesting questions. Why did echinoids, a small class during most of the Paleozoic, survive the Permo-Triassic boundary while many larger and more successful Paleozoic groups did not? I would conclude that small groups with advanced, newly-evolved morphology have perhaps the best chance of surviving a mass extinction in contrast to large successful groups (such as fusulinids or rugose corals) or groups that were relicts (such as trilobites) from earlier in the record.

When did the new types of regular echinoids and the vastly different irregular echinoids first originate? Durham & Melville (1957) took the fossil record as represented and indicated the times of origin as Late Triassic and Early Jurassic, respectively; Kier (1973, 1974, 1977b) agreed with this. In contrast, Durham (1966a, b) extended the time of origin back to the Mississippian and Pennsylvanian in the Late Paleozoic, respectively, thus assuming over 120 million years with no preserved fossil record. This seems unlikely because of Kier's recent work on the earliest known irregular echinoids in the Early Jurassic (Kier, unpublished), which show many features that are intermediate or just beginning to change. Also, if irregular echinoids had been present in the Late Paleozoic, they would have been even more likely to be preserved than regular cidaroids (Kier 1977a).

Why was so long a period (about 50 million years) required for the very successful irregular design to appear after regular echinoids survived the Permo-Triassic boundary? I would conclude that a period of time was necessary for echinoids to increase their diversity from the extremely low level at the beginning of the Triassic and that more advanced regular echinoids needed to evolve and diversify before the first irregular echinoids could evolve from them. Several other major groups in the Mesozoic (such as scleractinian corals) also show a similar delay before their first appearance.

4 ACKNOWLEDGMENTS

I thank the Geology Foundation, University of Texas at Austin, for research support during the preparation of this manuscript.

5 REFERENCES

Beerbower, J.R. 1968. *Search for the Past* (2nd ed.) 512pp. Prentice-Hall, Englewood Cliffs (N.J.).

Bell, B.M. 1976. A study of North American Edrioasteroidea. *Mem. N Y. State Mus. Sci. Serv.* 21:447pp.

Bell, B.M. 1977. Respiratory schemes in the Class Edrioasteroidea. *J. Paleontol.* 51:619-632.

Bell, B.M. & J.Sprinkle 1978. *Totiglobus,* an unusual new edrioasteroid from the Middle Cambrian of Nevada. *J. Paleontol.* 52:243-266.

Breimer, A. & D.B.Macurda 1972. The phylogeny of the fissiculate blastoids. *Verhand. Kon. Ned. Akad. Wetensch.* (1)26(3):1-390.

Breimer, A. & G.Ubaghs 1974. A critical comment on the classification of the Pelmatozoan echinoderms. I and II. *Proc. Konink. Ned. Akad. Wetensch.* (B)77:398-417.

Broadhead, T.W. 1980. Range extension – Middle Ordovician blastoids (Echinodermata) from the southern Appalachians. *Geol. Soc. Am., Abstr. with Progr.* 12(4):172.

Conway Morris, S. & H.B.Whittington 1979. The animals of the Burgess Shale. *Scient. American* 241:122-133.

Derstler, K.L. 1975. Carpoid echinoderms from Pennsylvania. *Geol. Soc. Am., Abstr. with Progr.* 7(1):48.
Denison, R.H. 1971. The origin of the vertebrates: a critical evaluation of current theories. *Proc. North Am. Paleont. Conv.* Part H:1132-1146.
Durham, J.W.1966a. Classification. In: R.C.Moore (ed.), *Treatise on Invertebrate Paleontology, Part U, Echinodermata,* vol.3a, pp.U270-295. Geol. Soc. Am. and Univ. Kansas Press, New York and Lawrence.
Durham, J.W. 1966b. Evolution among the Echinoidea. *Biol. Rev.* 41:368-391.
Durham, J.W. 1967a. Notes on the Helicoplacoidea and early echinoderms. *J. Paleontol.* 41:97-102.
Durham, J.W. 1967b. The incompleteness of our knowledge of the fossil record. *J. Paleontol.* 31:242-272.
Eaton, T.H. 1970. The stem-tail problem and the ancestry of chordates. *J. Paleontol.* 44: 969-979.
Eldredge, N. & S.J.Gould 1972. Punctuated equilibria: an alternative to phyletic gradualism. In: T.J.M.Schopf (ed.), *Models in Paleobiology,* pp.82-115. Freeman, Cooper & Co., San Francisco.
Gill, E.D. & K.E.Caster 1960. Carpoid echinoderms from the Silurian and Devonian of Australia. *Bull. Am. Paleontol.* 41:1-71.
Gould, S.J. 1977. *Ontogeny and phylogeny.* 501pp. Harvard Univ. Press, Cambridge (Mass.).
Gould, S.J. & N.Eldredge 1977. Punctuated equilibria: the tempo and mode of evolution reconsidered. *Paleobiol.* 3:115-151.
Haugh, B.N. 1973. Water vascular system of the Crinoidea Camerata. *J. Paleontol.* 47:77-90.
Haugh, B.N. & B.M.Bell 1980. Fossilized viscera in primitive echinoderms. *Science* 209: 653-657.
Hyman, L.H. 1955. *The Invertebrates: Echinodermata.* 763pp. McGraw-Hill Book Co., New York.
Jaekel, O. 1900. Ueber Carpoidean, eine neue classe von Pelmatozoen. *Z. Deutsch Geol. Gesell.* 53:661-678.
Jeffries, R.P.S. 1967. Some fossil chordates with echinoderm affinities. *Symp. zool. Soc. London* 20:163-208.
Jeffries, R.P.S. 1968. The Subphylum Calcichordata (Jeffries 1967), primitive fossil chordates with echinoderm affinities. *Bull. Brit. Mus. (nat. Hist.), Geol.* 16:243-339.
Jeffries, R.P.S. 1979. The origin of chordates – a methodological essay. In: M.R.House (ed.), *The Origin of Major Invertebrate Groups,* pp.443-447. Academic Press, London.
Jeffries, R.P.S. & R.J.Prokop 1972. A new calcichordate from the Ordovician of Bohemia and its anatomy, adaptations, and relationships. *Biol. J. Linn. Soc.* 4(2):69-115.
Kesling, R.V. 1967. Edrioasteroid with unique shape from Mississippian strata of Alberta. *J.Paleontol.* 41:197-202.
Kier, P.M. 1965. Evolutionary trends in Paleozoic echinoids. *J. Paleontol.* 39:436-465.
Kier, P.M. 1973. The echinoderms and Permian-Triassic time. In: Alan Logan & L.V.Hills (eds), *The Permian and Triassic Systems and their mutual Boundary,* pp.622-629. (Mem. Can. Soc. Petrol. Geol.).
Kier, P.M. 1974. Evolutionary trends and their functional significance in the post-Paleozoic echinoids. *J. Paleontol.* 48(suppl.3):1-95.
Kier, P.M. 1977a. The poor fossil record of the regular echinoid. *Paleobiol.* 3:168-174.
Kier, P.M. 1977b. Triassic echinoids. *Smith. Contr. Paleobiol.* 10:88pp.
Macurda, D.B. 1975. The *Pentremites* (Blastoidea) of the Burlington Limestone (Mississippian). *J. Paleontol.* 49:346-373.
Mayr, E. 1969. *Principles of systematic Zoology.* McGraw-Hill Book Co., New York, 428pp.

Nichols, D. 1969. *Echinoderms* (4th ed.). Hutchinson Univ. Library, London. 192pp.

Parsley, R.L. & K.E.Caster 1975. Zoological affinities and functional morphology of the Mitrata (Echinodermata). *Geol. Soc. Amer., Abstr. with Progr.* 7(7):1225-1226.

Parsley, R.L. & L.W.Mintz 1975. North American Paracrinoidea (Ordovician: Paracrinozoa, New, Echinodermata). *Bull. Amer. Paleontol.* 68:115pp.

Paul, C.R.C. 1967a. The functional morphology and mode of life of the cystoid *Pleurocystites*, E.Billings, 1854. *Symp. zool. Soc. Lond.* 20:105-123.

Paul, C.R.C. 1967b. New Ordovician Bothriocidaridae from Girvan and a reinterpretation of *Bothriocidaris* Eichwald. *Palaeontol.* 10:525-541.

Paul, C.R.C. 1977. Evolution of primitive echinoderms. In: Anthony Hallam (ed.), *Patterns of Evolution as Illustrated by the Fossil Record,* pp.123-158. Elsevier Scientific Publishing Co., Amsterdam.

Paul, C.R.C. 1979. Early echinoderm radiation. In: M.R.House (ed.), *The Origin of Major Invertebrate Groups,* pp.415-434. Academic Press, London.

Raup, D.M. & S.M.Stanley 1978. *Principles of Paleontology* (2nd ed.), 481pp. W.H.Freeman & Co., San Francisco.

Regnéll, G. 1960. 'Intermediate' forms in Early Palaeozoic echinoderms. *21st Inter. Geol. Congress,* Pt.22, p.71-80.

Repetski, J.E. 1978. A fish from the Upper Cambrian of North America. *Science* 200:529-531.

Simpson, G.G. 1953. *The Major Features of Evolution.* 434pp. Columbia Univ. Press, New York.

Sprinkle, J. 1969. The early evolution of crinozoan and blastozoan echinoderms. *Spec. Paper Geol. Soc. Am.* 121:287-288.

Sprinkle, J. 1973a. Morphology and evolution of blastozoan echinoderms. *Spec. Publ. Mus. comp. Zool. Harvard,* 283pp.

Sprinkle, J. 1973b. *Tripatocrinus,* a new hybocrinid crinoid based on disarticulated plates from the Antelope Valley Limestone of Nevada and California. *J. Paleontol.* 47:861-882.

Sprinkle, J. 1976. Biostratigraphy and paleoecology of Cambrian echinoderms from the Rocky Mountains. *Geol. Stud. Brigham Young Univ.* 23:61-73.

Sprinkle, J. 1980a. An overview of the fossil record. In: T.W.Broadhead & J.A.Waters (eds), *Echinoderms, Notes for a short Course,* pp.15-26 Univ. Tenn. Dept. Geol. Sciences.

Sprinkle, J. 1980b. Origin of blastoids: new look at an old problem. *Geol. Soc. Am., Abst. with Progr.* 12(7):528.

Sprinkle, J. (ed.), 1982. *Echinoderm Faunas from the Bromide Formation (Middle Ordovician) of Oklahoma.* Univ. Kansas Paleontol. Contr., Monograph 1, 369 pp.

Sprinkle, J. & B.M.Bell 1978. Paedomorphosis in edrioasteroid echinoderms. *Paleobiol.* 4:82-88.

Sprinkle, J. & R.A.Robison 1978. Ctenocystoids. In: R.C.Moore & Curt Teichert (eds), *Treatise on Invertebrate Paleontology, Part T, Echinodermata,* vol. 2c, pp.T998-1002. Geol. Soc. Am. and Univ. Kansas Press, New York and Lawrence.

Thompson, D. & R.E.Plotnick 1981. Cambrian vertebrates: are they arthropods? *Geol. Soc. Am., Abstr. with Progr.* 13(7):566.

Ubaghs, G. 1960. Le Genre *Lingulocystis* Thoral (Echinodermata, Eocrinoidea), avec des remarques critiques sur la position systématique du Genre *Rhipidocystis* Jaekel. *Annls Paléont.* 46:81-116.

Ubaghs, G. 1968. Stylophora. In: R.C.Moore (ed.), *Treatise on Invertebrate Paleontology, Part S, Echinodermata* vol. 1b, pp.S495-S565. Geol. Soc. Am. and Univ. Kansas Press, New York and Lawrence.

Ubaghs, G. 1975. Early Paleozoic echinoderms. *Ann. Rev. earth planet. Sci.* 3:79-98.

Valentine, J.W. 1969. Patterns of taxonomic and ecological structure of the shelf benthos during Phanerozoic time. *Palaeontology* 12:684-709.

Ziegler, A.M., C.R.Scotese, W.S.McKerrow, M.E.Johnson & R.K.Bambach 1979. Paleozoic paleogeography. *Ann. Rev. earth planet. Sci.* 7:473-502.

PHENOTYPIC VARIABILITY IN ECHINODERMS

NANCY H.MARCUS
Woods Hole Oceanographic Institution, Woods Hole, Mass., USA

CONTENTS

1 INTRODUCTION

The phylum Echinodermata has been one of the most widely recognized and studied group of invertebrates since ancient times. The group is mostly marine with some estuarine representatives. Individuals living in either an attached or free state can be found in all seas, from boreal to tropical latitudes, and from intertidal to abyssal depth zones. Echinoderms have a long fossil record that dates back to the early Cambrian. Today the phylum consists of five classes (Crinoidea, Holothuroidea, Echinoidea, Asteroidea, and Ophiuroidea). Excellent and exhaustive descriptive accounts of members of these groups are provided by Agassiz (1872-1874), H.L.Clark (1907), Mortensen (1928-1951), A.M.Clark (1962), Pawson (1965), and A.M.Clark & Rowe (1971). Some of the more recent reviews of echinoderm biology, ecology, and paleontology are those of Cuénot (1948), Hyman (1955), Harvey (1956), Boolootian (1966), R.Moore (1966, 1967), R.Moore & Teichert (1978), and Jangoux & Lawrence (1982). It is immediately apparent from these monographs and reviews that extant and extinct echinoderms represent a tremendously diverse group of animals.

The phenotype of an individual is the end product of a highly complex set of interactive events between the genotype and environment. It represents not only morphological attributes, but physiological, biochemical, and behavioral ones as well. A comprehensive account of phenotypic variability in echinoderms would essentially be a listing of most of the modern taxonomic literature on this group. Rather than cataloguing this variability, I will focus on those investigations which have attempted to elucidate the adaptive significance or cause of intraspecific phenotypic variation by either a thorough comparative field study or by experimental manipulation of individuals in the laboratory. Species representing each of the five classes of echinoderms have been studied in this regard, but the majority of the work has been conducted on echinoids. Moreover, most of these investigations have focused on morphological variability. Comparatively there is little information on intraspecific variation in physiological, biochemical (other than DNA or protein) or behavioral traits.

Many species are composed of populations that are morphologically distinct from one another. The differences expressed by individuals in these populations are the result of variation in their respective environments. However, it is not always obvious whether the differences are due to genetic variation and/or environmental modification of the phenotype. In most instances it is assumed that the differences reflect adaptation to the local environment, and although some morphological variation may not appear to be adaptively significant it may in fact represent the by-product of a physiologically adaptive process.

Evaluating the basis of phenotypic variation is a fundamental problem for the biologist who is interested in the proper phylogenetic classification of individuals. Systematists who study phenotypic variation are not just concerned with its occurrence, but more importantly with understanding its evolutionary significance, determining the relative contributions of the genotype and environment in regulating its expression, and the elucidation of how new species arise. The interpretation of phenotypic change has thus provided a basis for the formulation of many evolutionary theories and hypotheses concerning speciation.

Evolution has been defined as change in the genetic composition of populations. Since the phenotype is the result of interactions between the genotype and environment, genetic change is often manifest as variations in the phenotype. But all genetic change is not necessarily reflected by differences in the phenotype, and phenotypic variation is not always an accurate indicator of genetic variation. Ecotypes are races of varying distinctness which owe their most conspicuous characters to the selective effects of local environments (Mayr 1963). The genetic composition of ecotypes depends on their phenotypic flexibility which is ultimately determined by the genotype. Thus environmental factors may induce modification of phenotypes giving rise to phenoecotypes, or they may act as a sieve effecting the differential mortality of specific genotypes, giving rise to genecotypes. Phenoecotypes represent non-genetic adaptation which nevertheless is selectively adaptive. Different phenoecotypes are derived from similar genetic backgrounds, whereas genecotypes represent different genetic backgrounds.

Speciation is a result of the evolutionary process. It represents the acquisition of reproductive isolation and ecological compatibility of populations (Mayr 1954). Although the biological species concept is based on the premise that species are reproductively isolated units it is often difficult to demonstrate that two groups do not interbreed. However, because such isolation implies genetic differentiation, and therefore some degree of phenotypic differentiation, this latter property is usually studied to infer species distinctness. The separation of populations into geographic isolates is generally viewed as an important first step in the formation of new species, because it may result in the genetic divergence of populations, and thus eventually their reproductive isolation. That geographic speciation has been fundamental in the evolutionary history of some echinoderms was examined by Mayr (1954) for a group of tropical, littoral echinoids. Mayr based his analysis primarily on the classification scheme of Mortensen (1928-1951). Although Mayr suggested that some of the allopatric forms described by Mortensen did not necessarily warrant species status, he interpreted the distribution patterns of the different forms or species as evidence supporting the importance of geographic separation in the evolution and speciation of this group of echinoderms.

Until recently the approach used by most investigators to elucidate the evolutionary importance of phenotypic variation, was to correlate observed phenotypic change with parallel changes in environmental parameters (e.g. temperature, salinity, depth, substrate). This type of analysis provides some indication of the adaptive significance of the phenotypic variation, but it cannot provide insight into the relative contributions of the genotype and environment in regulating specific aspects of the variation. On the other hand by rearing echinoderms to maturity under controlled conditions in the laboratory, as has been done for several species, it is now possible to evaluate the genotypic and environmental components of some of the phenotypic variation in these animals. Gaining insight into the causes and adaptive significance of phenotypic variation expressed by individuals is fundamental to elucidating the evolutionary process. Such information is important to taxonomists, geneticists, ecologists, physiologists, paleontologists, and all biologists who deal with extant and extinct species.

2 MORPHOLOGICAL VARIATION

2.1 *Echinoidea*

Interpopulation morphological variation has been studied for a number of regular echinoids. Some of the characters that have been analyzed and compared are test shape, test thickness, number of spines, spine morphology, number of pore pairs, size of the apical system, arrangement of the ocular and genital plates, and color. In general the extent to which populations of a species are similar morphologically is more dependent on the habitat they occupy than on the geographic distance separating them. Thus populations from nearby localities, but occupying different habitats typically reflect greater differentiation than populations from distantly separated localities that are comparable ecological habitats.

Test shape, described by the ratio test height:test diameter, varies between populations which inhabit wave exposed and protected embayments. Lindhal & Runnström (1929) reported that S-type *Psammechinus miliaris* dredged from 40-58 m had smaller test height:test diameter ratios than Z-type individuals from *Zostera* beds. For some species it appears that flatter tests are more common among individuals from exposed sites with considerable wave action than from protected bays with calm water. This trend is evident for *Echinus esculentus* (H.B.Moore 1935), *Tripneustes ventricosus* (McPherson 1965) and *Echinometra lacunter* (Doran 1968). On the other hand, Dix (1970) reported no significant difference in test shape for populations of *Evechinus chloroticus* from such habitats, and McPherson (1968) reported that flatter tests of *Eucidaris tribuloides* were more common in calm areas. D'Arcy Thompson (1917) suggested that a flatter test permits greater contact of the tube feet with the substrate. Thus, H.B.Moore (1935) and McPherson (1965) postulated that a flatter test represents an adaptive response to intense wave action, because the increased number of tube feet in contact with the substrate should help to anchor the individual more securely in position. It would be possible to gain some insight into the selective advantage of test shape by investigating the behavior and survival of individuals transplanted from calm water to an exposed surf-beaten site. However, test shape variability need not be the only means of adapting to wave surge. It may be that a heavier test is adaptive in areas of intense wave action and has evolved as an alternative response for some species (see below). Another possible adaptation might be that the hydraulic properties of the water vascular system of some species are sufficient to generate enough suction power to anchor an individual in place in a variety of habitats regardless of test shape. Thus for *E.chloroticus* adaptation to wave action might not necessitate or require a change of test shape.

Spine length also varies between populations from areas differing in the degree of wave exposure. Short, thick spines with round tips prevail among individuals of *E.chloroticus* from surf beaten regions whereas long, thin spines with sharp tips are more typical of individuals from calm waters (Dix 1970). Dix did not discuss the adaptive significance of these differences, but it may be that long spines are a disadvantage to individuals occupying exposed areas inasmuch as such spines would probably be damaged more often due to the physical force of wave surge. The report by Ebert (1968) on *Strongylocentrotus pur-*

puratus indicates that the continual regeneration of broken spines utilizes energy that could otherwise be diverted to test growth or reproduction. In fact spine breakage and regeneration were more evident in specimens of *E.chloroticus* from the wave-exposed areas. Spine length differences also typify populations of *Arbacia punctulata* from Woods Hole, Massachusetts and Florida (Jackson 1927, Marcus 1980). By rearing individuals under the same conditions in the laboratory Marcus (1980) demonstrated that the long spines which are typical of individuals from the northeastern Gulf of Mexico and the short spines which are characteristic of individuals from Woods Hole are primarily due to genetic differences between the populations. However, other factors such as temperature may secondarily modify spine length by influencing their growth rate (Davies *et al.* 1972). The long spines of Florida individuals may have evolved as a defense against predators (Marcus 1980), or as suggested for *Echinostrephus molaris* (Campbell *et al.* 1973) as an efficient structure for trapping drift algae for food.

For individuals of *E.chloroticus* of similar test diameter, test thickness differs between populations from different habitats. Dix (1970) examined individuals from Kaikoura and Kaiteriteri, New Zealand; localities that differ in water temperature, wave action and food abundance. Test thickness was greater among individuals from the colder, wave exposed, and nutritionally rich area of Kaikoura. By maintaining individuals from both areas in the laboratory under two different feeding regimes Dix (1970) demonstrated that relative test thickness was significantly greater for fed than starved individuals after a six-month period. H.B.Moore (1935) observed differences in test thickness for individuals of *Echinus esculentus* from different habitats, and suggested that thicker tests are selectively favored in areas exposed to intense wave action. Doran (1968) reported that individuals of *Echinometra lacunter* from wave exposed sites in Barbados possessed thicker tests than individuals from calm waters. Individuals of *Strongylocentrotus droebachiensis* from wave-exposed sites are larger and the test is composed of a greater number of plates than individuals from calm water areas (Swan 1966). Swan suggested that these individuals would also be heavier, but actual weights were not determined. Heavier tests may represent an adaptation to intense wave action because they should be more resistant to physical displacement. Individuals of *Arbacia punctulata* from Woods Hole are heavier than individuals of similar test diameter from more southern areas (Marcus 1976). The differences may reflect different growth rates and/or rates of calcium deposition due to the different temperature (Weber 1973) or nutrient regimes experienced at these latitudes.

A number of other phenotypic characters differ among individuals of regular echinoid species from different regions. The weight of the Aristotle's lantern differs for populations of *Echinus esculentus* (Hagström & Lönning 1964), *Diadema setosum*, and *Strongylocentrotus purpuratus* (Ebert 1980). Ebert

(1980) reported for the latter two species that individuals with larger lanterns occurred in food-limited areas. He suggested that increased lantern size was due to a greater allocation of energy to the growth of this structure and was adaptive in areas where food was sparse because it enhanced the food gathering capabilities of the individual. It would be of interest to know if the tests of these individuals are not as heavy as ones from areas of abundant food, inasmuch as Dix (1970) reported a positive relationship between food abundance and test thickness. There may be a balance between test growth and lantern growth.

Variation in color pattern occurs in *Diadema antillarum* (Millott 1953, Kristensen 1964). Individuals with black spines are more common in quiet, shallow, clear water whereas individuals with grey, white or banded spines prevail in deep water or in surf-beaten regions under rocks. Young banded *D.antillarum* change spine color in response to ambient light conditions. Under normal light intensity individuals turn black in one to two months, but in darkness no color change occurs. Color differences also distinguish populations of *Psammechinus miliaris* (Lindhal & Runnström 1929), *Echinus esculentus* (H.B.Moore 1935), *Lytechinus variegatus* (H.B.Moore *et al.* 1963, Serafy 1973), *Arbacia punctulata* (Jackson 1912, 1927, H.L.Clark 1923), and *Echinometra mathaei* (Lawrence, in press). Some populations of *Eucidaris tribuloides* differ in the size of the peristome, and apical system (McPherson 1968). Interpopulation differentiation of the number of pore pairs occurs in *S.droebachiensis* (Swan 1958) and *Echinometra lacunter* (Doran 1968). For most echinoids, egg diameter is constant, however some species such as *Psammechinus miliaris* express considerable egg size variation (Lindahl & Runnström 1929). Lönning & Wennerberg (1963) suggested that egg size variation might depend on nutritional factors, but indicated that the genetic composition of the species might also be important.

A character that has received a great deal of attention by echinoid biologists is the development and arrangement of the ocular and genital plates that surround the periproct. Following metamorphosis all ocular plates are excluded from abutting the periproct. As development and skeletal growth proceeds the arrangement of the ocular plates may change so that some, all, or none eventually abutt (i.e. are inserted into) the periproct. Although Jackson's data (1912) suggest that the sequence of rearrangement is highly characteristic of a species, it is also evident that there is considerable interpopulation variation within a species (Jackson 1912, 1914, 1927). The data for *Strongylocentrotus droebachiensis* (Jackson 1912, Vasseur 1952, Swan 1966) indicate that individuals having a greater number of ocular plates insert correlates with increased salinities and colder temperatures. A trend associated with depth is also evident, which may be due to associated gradients of temperature and/or pressure. Geographic differentiation of ocular and genital plate arrangements also typify *Cidaris affinis, Arbacia punctulata, Tripneustes esculentus* and *Echinometra lacunter* (Jackson 1912, 1927). These studies do not prove that these physical factors are causal, nor do they shed light on the possible adaptive significance of the differences. Moreover, it is unclear whether the differences that arise are due to modifications induced by environmental factors during development or whether they actually reflect genetically determined growth rate differences. Environmental factors might produce phenocopies of genetically determined plate arrangements by imposing a differential effect on the rates at which rearrangements proceed. The data are intriguing though and as suggested by Swan (1966) warrant more careful investigation.

Intraspecific variation has been studied in some species of irregular echinoids. Differences in body shape, tuberculation, and color in *Meoma ventricosa ventricosa* correlate with water depth (Chesher 1970). Populations of *Brissopsis atlantica* from Colombia, South America and Florida, United States are distinct morphologically according to Chesher (1968), but it was not determined whether this was related to differences in their respective habitats. The functional significance of differences in body size and shape of some burrowing echinoids was thoroughly discussed by Nichols (1959, 1962). He suggested that test morphology and sediment texture influence the dimensions and structure of an individual's burrow, and therefore the morphology of individuals that can exist in the area. The same trend of body size and shape with sediment type has been documented for *Echinocardium cordatum, Spatangus purpureus,* and *Spatangus raschi,* as well as the extinct genus *Micraster* (Nichols 1959, 1962, Higgins 1974, 1975). The pattern typifies variation both within species and between species. Higgins (1974, 1975) pointed out that morphological differentiation among the three species of *Echinocardium* correlates better with the habitat than with geographic distance. Thus, individuals from distant areas (e.g. British Isles and New Zealand), but similar sediment types are morphologically more similar than individuals from nearby, but different ecological habitats. Scurry (1976) also reported that morphological variation of the extinct species *Nucleolites scutatus* correlates better with differences in the grain size of the geologic deposits, than with geographic distance between groups. The actual basis for the differences in test morphology are not known, but it appears that this character might not be an accurate indicator of genetic divergence.

A few studies report on phenotypic variation among sand dollars. As shown for the other echinoid groups morphological variation correlates better with features of the habitat rather than with the geographic distance separating populations. Variation in size and eccentricity of the apical system was studied in the genus *Dendraster* (Raup 1956, 1958, Stanton *et al.* 1979, Merrill & Hobson 1970). The character, eccentricity, refers to the position of the center of the apical system relative to the mid-point of the anterior-posterior axis of the test (Kew 1920). The degree of eccentricity has been one of the major characters used to classify extant and extinct members of the genus *Dendraster. Dendraster excentricus* occurs from Baja California north to British Columbia. Eccentricity values show a better correlation with environmental conditions than with geographic proximity (Raup 1956, 1968). Large and highly eccentric forms are more common along outer coast regions, whereas individuals in sheltered embayments tend to be smaller and less eccentric. These populations may be in very close geographic proximity and thus Raup (1956) believed they are ecotypes representing a phenotypic response to the environment and not evolutionary divergence. It was suggested that these differences reflect adaptation to water turbulence and sediment stability. *Dendraster* feeds in an inclined position by burying its anterior end in the substrate. Thus in turbulent areas individuals that are large and highly eccentric may be anchored more firmly in the sediment and better able to feed than smaller, less eccentric forms. It is less clear why highly eccentric forms are not also found in calm waters, but Stanton *et al.* (1979) suggested that the optimal form for *Dendraster* is reduced eccentricity and therefore the highly eccentric individuals only survive in conditions of rough water. Although no experimental evidence was cited, all of these investigators concluded that the observed morphological differences were due to environmental modification of the phenotype and not genetic differentiation. Moreover, none of these investigators supported the conclusion of MacGinitie & MacGinitie (1968) that the two forms are distinct subspecies.

Other morphological features of sand dollars that correlate with ecological conditions were reported by Lohavanijaya (1965) for *Echinarachnius parma* from the New England coast and by Raup (1958) for *D.excentricus* from the American west coast. A comparison of individuals from areas with different temperature regimes indicated that heavier tests are more common in colder regions. In *Dendraster excentricus* heavier tests are also associated with areas of intense wave action. The linear dimensions of *Echinarachnius parma* correlate with water movement. In flowing water the test tends to be longer than wide whereas in surf-swept beaches individuals are wider than long.

2.2 *Asteroidea, Ophiuroidea, Holothuroidea and Crinoidea*

Morphological differentiation of asteroids has been studied in a number of species (Tortonese 1943, 1950, Perlmutter & Nigrelli 1960, Pearse 1965, Achituv 1973, Strong 1975, Blake 1976, Lucas & Jones 1976, Franz *et al.* 1981). In the Mediterranean Sea, *Asterina wega* exists as several distinct populations that can be distinguished on the basis of ray size. The populations do not appear to be geographically isolated and Achituv (1973) suggested that the differences are environmentally induced. Two purportedly distinct species of *Linckia* are found in the coral reef community off Guam. They occupy different habitats and can be distinguished by ray size and shape, and color. Cross fertilization experiments in the laboratory were 100% successful with the two forms which led Strong (1975) to suggest that they be considered as ecotypes rather than two species. Individuals of the Antarctic asteroid, *Odontaster validus* from different localities manifest different coloration, which Pearse (1965) ascribed to dietary differences. The taxonomic status of two purportedly distinct species of *Acanthaster* was examined by rearing individuals in the laboratory (Lucas & Jones 1976). The results of the study indicate that the morphological features which had been used to distinguish *A.planci* and *A.brevispinus* are due to genetic differentiation. Despite the fact that the two forms were successfully crossed in the laboratory, hybrids are not found in the Great Barrier Reef region. Under natural conditions in the field individuals occupy different habitats and manifest different behaviors. Their designated specific status was supported by the study. A high incidence of four-rayed individuals results from development of *Echinaster* sp. in high salinity; temperature does not affect ray number (Watts *et al.* 1981).

Color differences distinguish populations of the ophiuroids, *Ophiocomomina nigra* (Fontaine 1962), *Amphipholis squamata* (Binaux & Bocquet 1971) and *Ophiothrix angulata* (Stancyk & Shaffer 1977). The differences shown by *O. nigra* were correlated with water depth and substrate. The color patterns of *A. squamata* were ascribed to genetic differences that were due to and maintained by self-fertilization. A suite of 14 morphological characters was examined by Serafy (1971) for a study of interpopulation variation in *Ophiopholis aculeata.* Some characters did not vary, others varied clinally, and some varied but without an apparent geographic trend. Considering all of the characters together some of the populations could be distinguished on a phenotypic basis, but the differences were not great enough to warrant the designation of subspecies.

Variation in the form, size and numerical density of the calcareous spicules occurring in the body wall of holothuroids has been fundamental to the designation of different species. Despite the fact that these spicules express a great deal of diversity (Filice 1950) they have nevertheless figured prominently in the classification of species from widely separated areas (H.L Clark 1907, Heding 1928, Madsen 1942). Swan (1966) suggested that the development of holothuroid spicules might be modified by environmental factors, and therefore not useful as accurate indicators of genetic divergence. This hypothesis is supported by the recent examination of spicule variation and genetic variation (detected by gel electrophoresis) in *Cucumaria curata.* Rutherford (1977) reported that this species expresses considerable spicule diversity along the coast of California yet shows little differentiation at the isozyme level. Spicule variation is related to ecological factors rather than geographic distance, but it is not clear whether this is due to genetic differentiation or environmental modification of the phenotype. The fact that genetic distance based upon electrophoretic analysis is small does not prove that the variation in spicule characteristics is non-genetic. This can only be demonstrated by carefully controlled breeding experiments. However, if spicule morphology is to be useful in taxonomic studies the influence of environmental factors on their appearance must be clarified.

Color differences distinguish populations of *Stichopus japonicus* (Choe & Ohshima 1961). The two prominent variants, red and green occupy different ecological habitats. The green form is more common on sandy bottoms, and areas of lower salinity whereas the red form predominates on gravel or pebble substrate, and in areas of higher salinity. The two types are also distinguished by a gelatinous coat that surrounds the egg of the red form, a layer absent in the egg of the green form. Slight but consistent morphological variation was observed for *Thyonella gemmata* (Manwell & Baker 1963) and it was shown by electrophoretic analysis that the slightly different morphological forms were genetically distinct. The different forms have been interpreted as sibling species.

Among the crinoids, Tortonese (1976) reported on geographic variation between populations of *Antedon* spp. in the Mediterranean. Although *Antedon mediterranea* is considered to be the representative species in this region, close morphological examination of the number of segments/cirri revealed that other species may be present.

2.3 *Variation in extinct genera*

Gaining an improved understanding of the factors that influence phenotypic variation among extant echinoderms provides information that is valuable for the interpretation of paleoenvironmental conditions and the phylogenetic status of fossil forms. The character of eccentricity has figured prominently in the classification of the genus *Dendraster.* The three fossil forms *D.ashleyi, D. gibbsii* and *D.gibbsii hamilis* were distinguished primarily on the basis of eccentricity and relative test thickness (Kew 1920). However, the results of several studies (Raup 1956, 1958, Stanton *et al.* 1979) of the extant species *Dendraster excentricus* suggest that these morphological features are not necessarily accurate indicators of genetic divergence. The range of eccentricity values expressed by the single species *Dendraster excentricus* is greater than the combined range of values representing *D.ashleyi, D.gibbsii,* and *D.gibbsii hamilis.* Similarly the extinct species *D.coalingaensis* has been distinguished from *D. gibbsii* on the basis of internal support structure, test thickness, and tuberculation. These morphological features also differ between individuals of *D.excentricus* from cold and warm water habitats, yet for this extant species such variation has been interpreted simply as phenoecotypic and insufficient to warrant separate species status. The information derived from the study of present-day species does not prove that the fossil forms are a single species, but it does suggest that some of the forms may simply represent phenoecotypic variation. Thus, a review of phenotypic variation among representatives of the extinct Pliocene genus *Merriamaster* and their geographic distribution led Durham

(1978) to conclude that the classification of *Merriamaster arnoldi* and *Merriamaster perrini* as two species was incorrect and that they should be considered instead as one polymorphic species.

The correlation between some morphological features and environmental conditions may be useful in the elucidation of paleoecological conditions. Test thickness, weight, and surface ornamentation of test plates are strongly correlated with temperature (Raup 1956, 1958, Oldfield 1976) and thus may be good indicators of different climatic conditions within a limited geographic region. Similarly, the test shape of burrowing animals, such as the heart urchins correlates with sediment type (Nichols 1959, 1962), and therefore, the test shape of extinct forms (Scurry 1976) should provide insight into the sedimentary characteristics of their habitat. The information derived from the study of these extant and fossil organisms is thus fundamental for the accurate and complete reconstruction of past biological communities and their physical environments.

3 PHYSIOLOGICAL VARIATION

Other phenotypic characters have been studied that relate to physiological differences (e.g. growth rate, reproductive periodicity, and salinity tolerance) between populations of echinoderms. Differences in the timing of reproductive activity of echinoids, asteroids, and ophiuroids have been reported for populations from different geographic localities (Pearse 1965, Sastry 1975, Hendler 1979, Stancyk & Shaffer 1977). These differences were ascribed to temperature, photoperiod and food differences. Although, in most cases it is not clear whether such variation is due to genetic differences, Sastry's (1975) investigation of *Arbacia punctulata* from Massachusetts and North Carolina suggests that the different reproductive cycles are due to non-genetic adaptation. Individuals transplanted from Massachusetts to North Carolina became acclimatized to the foreign environmental conditions and manifested identical cycles.

The salinity tolerance of two species from different areas has been investigated. Individuals of the ophiuroid, *Ophiothrix angulata* from Cedar Key, Florida were able to tolerate a wider range of salinity in the laboratory than ones from North Inlet, South Carolina (Stancyk & Shaffer 1977). Similarly, *Asterias rubens* from the Baltic Sea tolerates much lower salinities (8 %) than *A.rubens* from the North Sea (23 %) (Binyon 1961). The physiological differences expressed by individuals in the laboratory probably reflect adaptation to the different natural salinity regimes encountered in their respective habitats and may be due to genetic differentiation. Binyon (1961) suggested that the two populations of *A.rubens* were distinct physiological races. Watts *et al.* (1982) demonstrated that the development of two morphs of *Echinaster* sp. was affected differently by temperature and salinity levels. Crosses between the two morphs indicated that these differences were genetically based. Gezelius (1962) found that the salinity optima for development of embryos of *Psammechinus miliaris* depended upon the salinity at which the parents lived. His study showed acclimatization of populations, not adaptation. A genetic basis for differences in salinity tolerance has been demonstrated for oysters (Newkirk *et al.* 1977).

Intraspecific growth rate variation has been studied in a number of species (see review by Swan 1966), but there is little experimental evidence to indicate whether the differences are due to genetic factors. Some investigators compared growth rates in the field by measuring tagged or caged individuals. For example, the growth rate of *S.droebachiensis* from the Pacific coast of Washington differs from animals from the Maine-New Hampshire coast (Swan 1961). The principal component of the diet of the American west coast populations is *Nereocystis* and of the American east coast populations, *Laminaria.* Thus the growth rate variation may be due to these food differences (see Lawrence and Lane, 1982), as well as other environmental factors (e.g. temperature, salinity). Other investigators examined growth rate variation under laboratory conditions, but few analyzed interpopulation differences. In several investigations, actual growth was not measured, but rather inferred from a size-frequency analysis of populations. Off southern Florida the test size of *Echinometra lacunter* was generally larger for individuals from inshore populations than offshore populations (McPherson 1969). These differences may reflect different food levels and/or growth rates. Similarly, the size range of *A.punctulata* from the Gulf of Mexico was greater than the size range of the population in Woods Hole (Marcus 1976). These differences might also be due to growth rate differences. On the other hand, southern individuals might survive longer than northern individuals. If such individuals continued to grow they would attain a larger size. If the populations are in close geographic proximity, as in the case of *E.lacunter* (McPherson 1969), it seems unlikely that the growth rate differences are due to genetic divergence since gene flow probably occurs between the populations. However, for populations from widely separated areas (e.g. *S.droebachiensis* and *A.punctulata)* genetic differentiation of growth rates may occur as has been shown for individuals of the gastropod *Crepidula fornicata* from Massachusetts and Delaware (Ament 1977).

4 CAUSE AND SIGNIFICANCE OF PHENOTYPIC VARIATION

All species of echinoderms express some degree of phenotypic variability, whether it is related to sex, age, or ecological differences. It is not clearly understood, however, why some species express a greater degree of phenotypic variability than others. It is evident from the study of echinoderms that characters which are highly variable in one species (e.g. test shape) may vary very little in another species. The extent to which a character is variable may be indicative of its contribution to the overall fitness of a species. Harrison (1959) suggested that highly variable characters have little impact on fitness and are thus free to vary. On the other hand, highly variable characters may enable a species to exist in different habitats.

Phenotypic characters, primarily at the morphological level, have constituted the major criteria used by taxonomists to identify and classify individuals into species categories and higher taxa. The value of these characters as accurate indicators of phylogenetic relationships and genetic divergence is a function of their variability and the extent to which such variation is due to genetic and non-genetic factors. The use of morphological characters has proved to be sufficient for the accurate classification of most echinoderms. However, exceptions such as sibling species which cannot be distinguished at the morphological level are known to exist. During the last few decades electrophoretic analysis of protein variation has proved to be a valuable tool for the elucidation of genetic differentiation and phylogenetic relationships. The soluble proteins that are detected by this method represent the direct products of gene activity. It is assumed that variation in protein patterns due to environmental modification is minimal at this level of analysis. In general, conclusions originally derived from the study of morphological characters, concerning the evolutionary relationships of populations, have been substantiated by data obtained from electrophoretic analyses. However, the extent of evolutionary divergence of populations based on morphological and electrophoretic analyses is not always comparable. In a recent study, Lessios (1981) compared the morphological and molecular (enzyme) divergence of geminate echinoid species pairs from the east and west coasts of Panama. Although representatives of each pair were presumably isolated from one another at the same time due to the emergence of the Isthmus of Panama, it is apparent that the rates of divergence at the morphological and molecular levels have been different among the three genera (*Diadema, Echinometra* and *Eucidaris*) that were studied. Moreover, the extent of divergence at the two levels is not comparable for all species. For example, *Echinometra lacunter* and *E. viridis* exist sympatrically. At the molecular level they are quite similar, but they are quite distinct at the morphological level. Lessios suggested that the lack of congruence is due to isozyme variability being influenced by components of the environment that are different from those that influence morphological variation. The rates of morphological and molecular divergence are therefore judged to be independent of one another.

The planktonic larval stage of most echinoderms provides the major potential agent of gene flow between geographic populations. However, the fact that a larva can survive several weeks in the plankton, and be carried by currents to new areas does not insure that it will metamorphose, survive, and reproduce in the foreign location. The extent to which geographic populations are genetically adapted to their particular environmental regime will influence the survival of their offspring in other localities since different genetic combinations should be favored in different areas. The extent of genetic differentiation will depend therefore on a balance between geneflow due to larval transport and differential selection of specific genotypes. Species that reproduce by broadcasting larvae into the oceans at each spawning may increase the probability of successful colonization by producing developmentally plastic individuals. Thus, within certain environmental limits the morphology expressed by an individual might be variable and determined by the conditions experienced during development. Phenotypic plasticity might be more common to immobile species such as echinoderms that are unable to migrate and are thus forced to cope with environmental fluctuations (Heywood 1959). However, such phenoecotypic adaptation need not preclude genotypic adaptation which as Heywood (1959) suggested may be better suited to long-term adaptation. Phenotypic plasticity provides individuals with the capacity for immediate adaptation to environmental variation, and eventually through evolutionary time, genetic adaptation may arise due to mutation and differential selection. Thus for populations where genetic divergence based on electrophoretic differences appears to be greater than morphological differentiation developmental canalization may have restricted phenotypic variation (Nixon & Taylor 1977). On the other hand, for populations where morphological variation exceeds electrophoretically detectable protein divergence, phenotypic plasticity may be present (Gould *et al.* 1974, Marcus 1976, Rutherford 1977).

Many of the investigators who have examined the phenotypic diversity of echinoderms have discussed the relative importance of several evolutionary mechanisms in determining the variation expressed by populations from different habitats (Achituv 1973, Dix 1970, Lohvanijaya 1965, Marcus 1980, McPherson 1968, 1969, Merrill & Hobson 1970, Nichols 1962, Raup 1956, 1958, Stanton *et al.* 1979, Swan 1958, 1961, 1966). The two mechanisms considered most often are the differential selection of specific genotypes and environmentally induced modification of the phenotype. However, the data from these studies are insufficient to demonstrate the actual cause for the species examined.

Furthermore, although electrophoretic studies provide useful means to estimate the genetic divergence of populations, they are insufficient to distinguish genetic from environmentally induced causes of specific phenotypic characters. Only by rearing individuals under controlled conditions in the laboratory, will it be possible to distinguish between genetic and environmental factors controlling the phenotype. This approach has helped clarify causes of phenotypic variation in a few species of asteroids (Lucas & Jones 1976), and echinoids (Hinegardner 1975, Marcus 1980, Pawson & Miller 1982). When actual breeding programs cannot be established, laboratory studies that are designed to examine the response of the same individual or genetically identical individuals to changes in environmental parameters should be useful for elucidating the genetic and environmental components of physiological and behavioral traits. The excellent regenerative capacity of echinoderms makes them ideal subjects for such experimental analysis. An alternative approach would be to compare the phenotype of identical twins reared under similar (Marcus 1979) and different environmental regimes in the laboratory.

Elucidating the genetic and non-genetic factors that influence phenotypic variation is essential not only to accurately clarify the evolutionary relationships of echinoderm taxa, but also such information should provide a better understanding of how species cope with and adapt to environmental heterogeneity. Insight into the adaptive significance of specific characters should be gained by studying species that express similar trends in variation in relation to the environment. The study of phenotypic characters usually provides a good indication of phylogenetic status. However, the fact that some unrelated species reflect similar trends in morphological variation due to convergent evolution indicates that phylogenetic relationships might sometimes be difficult to interpret based upon these features alone. Moreover, it remains unclear as to what types of changes are essential to the formation of a new species. Recent evidence suggests that regulatory genes (Turner *et al.* 1979, Larson 1980) which control developmental processes and growth rates may be of primary importance.

5 REFERENCES

Achituv, Y. 1973. On the distribution and variability of the Indo-Pacific sea star *Asterina wega* (Echinodermata: Asteroidea) in the Mediterranean Sea. *Mar. Biol.*, 18:333-336.

Agassiz, A. 1872-1874. Revision of the Echini. *Mem. Mus. comp. Zool. Harv.*, 3:1-762.

Ament, A. 1979. Geographic variation in relation to life history in three species of the marine gastropod genus *Crepidula:* Growth rates of newly hatched larvae and juveniles. In: S.Stancyk (ed.), *Reproductive Ecology of Marine Invertebrates,* pp.61-76. Univ. South Carolina Press: Columbia.

Binaux, R. & C.Bocquet 1971. Sur le polychromatisme de l'ophiure *Amphipholis squamata* (Delle Chiaje). *C.R. Acad. Sc. Paris,* 273:1618-1619.

Binyon, J. 1961. Salinity tolerance and permeability to water of the starfish *Asterias rubens* L. *J. mar. biol. Ass. UK,* 41:161-174.

Blake, D.B. 1976. Sea star ossicle morphology: taxonomic implications. *Thalassia Jugosl.* 12:21-29.

Boolootian, R.A. (ed.) 1966. *Physiology of Echinodermata.* John Wiley & Sons, New York.

Campbell, A.J., Dart, S.Head & R.Ormond 1973. The feeding activity of *Echinostrephus molaris* (de Blainville) in the Central Red Sea. *Mar. behav. Physiol.* 2:155-169.

Chesher, R.H. 1968. The systematics of sympatric species in West Indian spatangoids: A revision of the genera *Brissopsis, Plethotaenia, Paleopneustes,* and *Savianaster. Stud. trop. Oceanogr. (Miami),* 7:viii+168 pp.

Chesher, R.H. 1970. Evolution in the genus *Meoma* (Echinoidea: Spatangoida) and a description of a new species from Panama. *Bull. mar. Sci.* 20:731-761.

Choe, S. & Y.Oshima 1961. On the morphological and ecological differences between two commercial forms, 'Green' and 'Red', of the Japanese common sea cucumber, *Stichopus japonicus* Selenka. *Bull. Jap. Soc. sci. Fish.* 27:97-106.

Clark, A.M. 1962. *Starfishes and their relations.* British Museum: London.

Clark, A.M. & F.W.Rowe 1971. *Monograph of shallow water Indo-West Pacific Echinoderms.* Publ. Trustees Brit. Mus. (Nat. Hist.), London.

Clark, H.L. 1907. The apodous holothurians. *Smithson. Cont. Knowl.* 35:1-231.

Clark, H.L. 1923. The distribution and derivation of some New England echinoderms. *Am. Nat.* 62:229-237.

Cuénot, L. 1948. Anatomie, éthologie et systématiques des échinodermes. In: P.Grasse (ed.), *Traité de Zoologie,* Vol.11, pp.3-363. Masson, Paris.

Davies, T., M.Crenshaw & B.Heatfield 1972. The effect of temperature on the chemistry and structure of echinoid spine regeneration. *J.Paleont.* 46:874-884.

Dix, T.G. 1970. Biology of *Evechinus chloroticus* (Echinoidea: Echinometridae) from different localities. *N.Z. Jl. Mar. Freshw. Res.* 4:91-116.

Doran, G. 1968. *Ecology of Echinometra lucunter (Linnaeus) a West Indian Echinoid.* M.Sc. Thesis, McGill University.

Durham, J.W. 1978. Polymorphism in the Pliocene sand dollar *Merriamaster* (Echinoidea). *J.Paleont.* 52:275-286.

Ebert, T.A. 1968. Growth rates of the sea urchin *Strongylocentrotus purpuratus* related to food availability and spine abrasion. *Ecology,* 49:1075-1091.

Ebert, T.A. 1980. Relative growth of sea urchin jaws: an example of plastic resource allocation. *Bull. mar. Sci.* 30:467-474.

Filice, F. 1950. A study of some variations in *Cucumaria curata* (Holothuroidea). *Wasmann J. Biol.* 8:39-47.

Fontaine, A.R. 1962. The colours of *Ophiocomina nigra* (Abildgaard). I. Color variation and its relation to distribution. *J. mar. biol. Ass. UK,* 42:1-8.

Franz, D.R., E.Worley & A.Merrill 1981. Distribution patterns of common seastars of the Middle Atlantic continental shelf of the Northwest Atlantic (Gulf of Maine to Cape Hatteras). *Biol. Bull.* 160:394-418.

Gezelius, G. 1962. Adaption of the sea urchin *Psammechinus miliaris* to different salinities. *Zool. Bidrag Uppsala* 35:329-337.

Gould, S., D.Woodruff & J.Martin 1974. Genetics and morphometrics of *Cerion* at Pongo Carpet: A new systematic approach to this enigmatic land snail. *Syst. Zool.* 23:518-535.

Hagström, B.E. & S.Lönning 1964. Morphological variation in *Echinus esculentus* from the Norwegian west coast. *Sarsia* 17:39-46.

Harrison, A. 1959. Environmental determination of the phenotype. *Publ. Syst. Assoc.* 3: 81-86.

Harvey, E.B. 1956. *The American Arbacia and Other Sea Urchins.* Princeton Univ. Press: Princeton.

Heding, S. 1928. Synaptidae. *Vidensk. medd. Dansk. naturh. Foren. Kbh.* 85:105-323.

Hendler, G. 1979. Reproductive periodicity of ophiuroids (Ophiuroidea: Echinodermata) on the Atlantic and Pacific coasts of Panama. In: S.Stancyk (ed.), *Reproductive Ecology of Marine Invertebrates,* pp.145-156. Univ. South Carolina Press, Columbia.

Heywood, V. 1959. The taxonomic treatment of ecotypic variation. *Publ. Syst. Assoc.* 3: 87-112.

Higgins, R.C. 1974. Specific status of *Echinocardium cordatum, E.australe,* and *E.zealandicum* (Echinoidea: Spatangoida) around New Zealand, with comments on the relation of morphological variation to the environment. *J. Zool., Lond.* 173:451-475.

Higgins, R.C. 1975. Observations on the morphology of *Echinocardium cordatum* (Echi-

noidea: Spatangoida) from diverse geographical areas. *J. Zool., Lond.* 177:507-515.
Hinegardner, R.T. 1975. Morphology and genetics of sea urchin development. *Am. Zool.* 15:679-689.
Hyman, L.H. 1955. *The Invertebrates. Vol. IX, Echinodermata.* McGraw-Hill, New York.
Jackson, R.T. 1912. Phylogeny of the Echini, with a revision of Paleozoic species. *Mem. Boston Soc. nat. Hist.* 5:89-153.
Jackson, R.T. 1914. Studies of Jamaica Echini. *Publs Carnegie Inst. Washington,* 183:139-162.
Jackson, R.T. 1927. Studies of *Arbacia punctulata* and its allies. *Mem. Boston Soc. nat. Hist.* 8:434-565.
Jangoux, M. & J.M.Lawrence (eds) 1982. *Echinoderm Nutrition.* 654 pp. Balkema, Rotterdam.
Kew, W. 1920. Cretaceous and Cenozoic Echinoidea of the Pacific coast of North America. *Publ. Univ. Calif. Geol. Sci.* 12:23-236.
Kristensen, I. 1964. The effect of raising *Diadema* at different levels of light intensity on pigmentation and preference for darkness. *Carribean J. Sci.* 4:441-443.
Larson, A. 1980. Paedomorphosis in relation to rates of morphological and molecular evolution in the salamander *Aneides flavipunctatus* (Amphibia, Plethadontidae). *Evolution* 34:1-17.
Lawrence, J.M. & J.M.Lane 1982. The utilization of nutrients by post-metamorphic echinoderms, pp. 331-337. In: M.Jangoux & J.M.Lawrence (eds), *Echinoderm Nutrition.* Rotterdam: Balkema.
Lawrence, J.M. (in press). Alternate states of populations of *Echinometra mathaei* (de Blainville) (Echinodermata: Echinoidea) in the Gulf of Suez and the Gulf of Aqaba. *Proc. intern. Red Sea Conf., Hurdaqua.*
Lessios, H.A. 1981. Divergence in allopatry: molecular and morphological differentiation between sea urchins separated by the Isthmus of Panama. *Evolution* 35:618-634.
Lindhal, P.E. & R.Runnström 1929. Variation und ökologie von *Psammechinus miliaris* (Gmelin). *Acta Zool.* 10:401-484.
Lohavanijaya, P. 1965. Variation in linear dimensions, test weight, and ambulacral pores in the sand dollar, *Echinarachnius parma* (Lamarck). *Biol. Bull.* 128:404-414.
Lönning, S. & C.Wennerberg 1963. Biometric studies of echinoderm eggs. *Sarsia* 11:25-27.
Lucas, J.S. & M.Jones 1976. Hybrid crown-of-thorns starfish (*Acanthaster planci* x *A. brevispinus*) reared to maturity in the laboratory. *Nature, London* 263:409-412.
Madsen, F.J. 1942. *Cucumaria hyndmani.* The variation of its calcareous deposits. *Vidensk. medd. Dansk. naturh. Foren. Kbh.* 105:395-406.
MacGinitie, G.E. & N.MacGinitie 1968. *Natural History of Marine Animals.* McGraw-Hill, New York.
Manwell, C. & C.Baker 1963. A sibling species of sea cucumber discovered by starch gel electrophoresis. *Comp. Biochem. Physiol.* 10:39-53.
Marcus, N. 1976. Genotypic and phenotypic variation in the sea urchin, *Arbacia punctulata.* Ph.D. Dissert., Yale University.
Marcus, N.H. 1979. Developmental aberations associated with twinning in laboratory-reared sea urchins. *Devel. Biol.* 70:274-277.
Marcus, N.H. 1980. Genetics of morphological variation in geographically distant populations of the sea urchin, *Arbacia punctulata* (Lamarck). *J. exp. mar. Biol. Ecol.* 43:121-130.
Mayr, E. 1954. Geographic speciation in tropical echinoids. *Evolution,* 8:1-8.
Mayr, E. 1963. *Animal Species and Evolution.* Belknap Press: Cambridge.
McPherson, B.F. 1965. Contributions to the biology of the sea urchin *Tripneustes ventricosus. Bull. mar. Sci.* 15:228-244.

McPherson, B.F. 1968. Contributions to the biology of the sea urchin *Eucidaris tribuloides* (Lamarck). *Bull. mar. Sci.* 18:400-443.
McPherson, B.F. 1969. Studies on the biology of the tropical sea urchins *Echinometra lucunter* and *Echinometra viridis. Bull. mar. Sci.* 19:194-213.
Merrill, R.J. & E.S.Hobson 1970. Field observations of *Dendraster excentricus,* a sand dollar of western North America. *Am. Mid. Nat.* 83:595-624.
Millott, N. 1953. Colour pattern and definition of the species, *Diadema antillarum* Philippi. *Experientia* 9:98-99.
Moore, H.B. 1935. A comparison of the biology of *Echinus esculentus* in different habitats. Part II. *J. mar. biol. Assoc. U.K.* 20:109-128.
Moore, H.B., T.Jutare, J.Bauer & J.Jones 1963. The biology of *Lytechinus variegatus. Bull. mar. Sci.* 13:25-53.
Moore, R.C. (ed.) 1966. *Treatise on Invertebrate Paleontology. Part U, Echinodermata 3.* Univ. Kansas: Lawrence.
Moore, R.C. (ed.) 1967. *Treatise on Invertebrate Paleontology. Part S, Echinodermata 1.* Univ. Kansas: Lawrence.
Moore, R.C. & C.Teichert (eds) 1978. *Treatise on Invertebrate Paleontology. Part T, Echinodermata 2.* Univ. Kansas: Lawrence.
Mortensen, T. 1928-1951. *A Monograph of the Echinoidea.* C.A.Reitzel: Copenhagen.
Newkirk, G., D.Waugh & L.Haley 1977. Genetics of larval tolerance to reduced salinities in two populations of oysters, *Crassostrea virginica. J. Fish. Res. Bd. Can.* 34:384-387.
Nichols, D. 1959. Mode of life and taxonomy in irregular sea urchins. *Publ. Syst. Assoc.* 3:61-80.
Nichols, D. 1962. Differential selection in populations of a heart-urchin. *Publ. Syst. Assoc.* 4:105-118.
Nixon, S. & R.Taylor 1977. Large genetic distances associated with little morphological variation in *Polycelis coronata* and *Dugesia tigrina* (Planaria). *Syst. Zool.* 26:152-164.
Oldfield, S.C. 1976. Surface ornamentation of the echinoid test and its ecologic significance. *Paleobiology* 2:122-130.
Pawson, D.L. 1965. Evolution and phylogeny of the Holothuroidea. In: R.C.Moore (ed.), *Treatise on invertebrate paleontology, Part U, Echinodermata 3.* Univ. Kansas, Lawrence.
Pawson, D.L. & J.E.Miller 1982. Studies of genetically controlled phenotypic characters in laboratory-reared *Lytechinus variegatus* (Lamarck) (Echinodermata: Echinoidea) from Bermuda and Florida. In: J.M.Lawrence (ed.), *Proc. 4th Intern. Echinoderm Conf., Tampa Bay.* pp. 165-171. Balkema, Rotterdam.
Pearse, J.S. 1965. Reproductive periodicities in several contrasting populations of *Odontaster validus* Koehler, a common Antarctic asteroid. *Antarctic Res.* 5:39-85.
Perlmutter, A. & R.Nigrelli 1960. A possible hybrid population of the starfish, *Asterias forbesi* (Desor) in western Long Island Sound. *Proc. natl Shellf. Assoc.* 51:93-96.
Raup, D.M. 1956. *Dendraster:* A problem in echinoid taxonomy. *J. Paleont.,* 30:685-694.
Raup, D.M. 1958. The relation between water temperature and morphology in *Dendraster. J. Geol.* 66:668-677.
Rutherford, J.C. 1977. Geographical variation in morphological and electrophoretic characters in the holothurian *Cucumaria curata. Mar. Biol.* 43:165-174.
Scurry, D.G.A. 1976. Variation within the upper Jurassic irregular echinoid *Nucleolites scutatus. Thalassia Jugosl.* 12:323-334.
Sastry, A. 1975. Physiology and ecology of reproduction in marine invertebrates. In: F.J. Vernberg (ed.), *Physiological Ecology of Estuarine Organisms.* Univ. South Carolina Press, Columbia.
Serafy, D.K. 1971. Intraspecific variation in the brittle-star *Ophiopholis aculeata* (Lin-

naeus) in the Northwestern Atlantic (Echinodermata, Ophiuroidea). *Biol. Bull.* 140: 323-330.

Serafy, D.K. 1973. Variation in the polytypic sea urchin *Lytechinus variegatus* (Lamarck 1816) in the western Atlantic (Echinodermata, Echinoidea). *Bull. mar. Sci.* 23:525-534.

Stancyk, S.G. & P.L.Shaffer 1977. The salinity tolerance of *Ophiothrix angulata* (Say) (Echinodermata: Ophiuroidea) in latitudinally separate populations. *J. exp. mar. Biol. Ecol.* 29:35-43.

Stanton, R.J., J.R.Dodd & R.R.Alexander 1979. Eccentricity in the clypeasteroid echinoid *Dendraster:* environmental significance and application in Pliocene paleoecology. *Lethaia,* 12:75-87.

Strong, R. 1975. Distribution, morphometry, and thermal stress studies on two forms of *Linckia* (Asteroidea) on Guam. *Micronesia* 11:167-183.

Swan, E.F. 1958. Growth and variation in sea urchins of York, Maine. *J. mar. Res.* 17: 505-522.

Swan, E.F. 1961. Some observations on the growth rate of sea urchins in the genus *Strongylocentrotus. Biol. Bull.* 120:420-427.

Swan, E.F. 1966. Growth, autotomy, and regeneration. In: R.Boolootian (ed.), *Physiology of the Echinodermata.* pp. 397-434. Interscience Publ.: New York.

Thompson, D'A. 1917. *On Growth and Form.* Cambridge University Press, Cambridge. 793pp.

Tortonese, E. 1943. Variazioni fenotipiche e biologia della popolazione di *Astropecten aranciacus* (echinodermi) del Golfo di Napoli, con riferimenti a specie congeneri. *Boll. Museo Zool. Univ. Torino,* 1:87-123.

Tortonese, E. 1950. Differenzione geographica ed ecologia negli asteroidi. *Boll. Museo Zool. Univ. Torino,* 17:339-354.

Tortonese, E. 1976. On the species of *Antedon* (Crinoidea) present along the coasts of Sicily. *Thalassia Jugosl.* 12:355-359.

Turner, J., M.Johnson & W.Eanes 1979. Contrasted modes of evolution in the same genome: allozymes and adaptive change in *Heliconius. Proc. natl Acad. Sci.* 76:1924-1928.

Vasseur, E. 1952. Geographic variation in the Norwegian sea urchins, *Strongylocentrotus droebachiensis* and *S.pallidus. Evolution* 6:87-100.

Watts, S.A., R.E.Scheibling, A.G.Marsh & J.B.McClintock 1982. Effect of temperature and salinity on larval development of two *Echinaster* (Echinodermata: Asteroidea) sibling species of morphs and their hybrids. *Biol. Bull.* 163: 348-354.

Weber, J.N. 1973. Temperature dependence of magnesium in echinoid and asteroid skeletal calcite: a reinterpretation of its significance. *J. Geol.* 81:543-556.

GENOMIC VARIABILITY IN ECHINODERMS

SYDNEY P.CRAIG
Department of Biological Sciences, Warren Wilson College, Sivannanoa, NC, USA

CONTENTS

1 INTRODUCTION

Echinoderms, particularly echinoids, have been the object of intensive analyses of genetic structure and evolutionary change. The reasons for this include the abundance of extant species, the ease of radiolabelling and extracting DNA,

Table 1. Estimates for repetitive sequence contents for the genomes of echinoderms.

Species	Approximate genome size+	Minimum percent of genome in repeat DNA	% of long or clustered repeats	% of short repeats	References
(A) Holothuroidea					
Stichopus japonicus	0.79 pg*	35-45 %	3.4-6.9 %**	12-23 %**	Brykov *et al.* 1979b
(B) Echinoidea					
Arbacia lixula	0.76 pg*	⩾32.0 %	24.5 %	⩾ 7.5 %	Craig (unpublished)
Sphaerechinus granularis	–	⩾31.7 %	16.1 %	⩾15.6 %	Craig (unpublished)
Lytechinus variegatus	0.92 pg	⩾26.8 %	21.7 %	⩾ 5.1 %	Craig (unpublished)
Paracentrotus lividus	–	20-33 %	13 %	7-20 %	Craig *et al.* 1979
Psammechinus microtuberculatus	–	⩾22.4 %	10 %	⩾12.4 %	Craig (unpublished)
Strongylocentrotus intermedius	0.83-0.9 pg*	35-40 %	3.4-6.9 %	12-23 %	Brykov *et al.* 1979b
Strongylocentrotus nudus	0.83-0.9 pg*	35-40 %	3.4-6.9 %	12-23 %	Brykov *et al.* 1979b
Strongylocentrotus purpuratus	0.89 pg	25 %	–	18 %	Davidson *et al.* 1975a
		19 %	6 %	12.9 %	Moore *et al.* 1981
(C) Asteroidea					
Dermasterias imbricata	0.54 pg	19.2-55 %	4-5 %	–	Smith *et al.* 1980
Asterias amurensis	0.6 pg*	30-42 %	3.4-6.9 %**	12-23 %**	Brykov *et al.* 1979b
Pisaster ochraceus	0.58-0.68 pg*	22.4 %	11.1 %	7.2 %	Smith & Boal 1978

The ranges of values listed above are generally the result of differences in S_1 nuclease used. Values may also vary due to S_1 digestion of short repeat sequences and to differences in Cot values at the termination of specific reassociations of DNA.

\+ All estimates for genome size come from Hinegardner 1974.

* If the genome size for a species has not been published, the size, or range of sizes, for species belonging to the same genus are listed.

** The ratio of long to short repetitive DNA is 0.33 for *A.amurensis* and 0.1 for *St.japonicus* (Brykov *et al.* 1979b).

and the availability of relatively extensive fossil records. Furthermore, phenotypic stability and constancy of DNA content, particularly among echinoids (Hinegardner 1974), tend to minimize potential variables that might interfere with efforts to determine the important details of genomic organization and change. Despite these advantages, some of the established dogma concerning the organization and evolution of specific components within the genomes of echinoids and other organisms have been found to be in error. This review considers the organization of DNA in echinoderms, evolutionary changes in DNA sequences, and the possible significance of rearrangement of the sequences during evolution.

2 CHARACTERIZATION OF THE GENOMES OF ECHINODERMS

2.1 *The general organization of eucaryotic DNA*

One of the significant differences between the organization of the DNA in most eucaryotes and bacteria is the relative abundance of repetitive (redundant or reiterated) sequences in the eucaryotic genomes (Britten & Kohne 1968; for a review see Lewin 1980). Although repetitive sequences may comprise more than 80% of the genome of some higher eucaryotes (e.g. certain plants, Flavell *et al.* 1980), the typical amount present is generally closer to 25-30% of the total DNA (Weinblum *et al.* 1973, Davidson *et al.* 1975a, Galau *et al.* 1976a; see Appendix). These repetitive sequences are organized within the genomes of most higher eucaryotes in one of two general patterns. The first pattern, the long period interspersion pattern, has repetitive sequences organized almost exclusively as arrays averaging more than two thousand base pairs (two kilobases) in length. These may be composed of clustered identical or clustered nonidentical repeats, involving repetitive sequences interspersed among sequences belonging to other 'families' of repetitive sequences (a family refers to all of the copies of a sequence that are similar enough to indicate that they are descendants of an ancestral sequence that was amplified in the evolutionary past). The second pattern, the short period interspersion pattern, contains short repetitive sequences with average lengths of 250-350 base pairs interspersed among the unique sequences (also referred to as non-repetitive or single copy DNA).

Organisms possessing the long period interspersion pattern have only very small quantities of short repetitive sequences interspersed with the unique DNA. However, organisms having the short period interspersion pattern, invariably also possess substantial quantities of long repetitive DNA. The long period pattern of sequence organization is occasionally referred to as the *Drosophila* pattern, whereas the short period interspersion pattern is sometimes referred to as the *Xenopus* pattern, after the organisms in which they were first des-

Table 2. Identified components in clustered repetitive DNA sequences of echinoids

Component	Species	Reiteration frequency	Repeat size+	% of genome	References
Genes for 26s and 18s rRNA	*Lytechinus variegatus*	170	11.4 kb	0.22 %*	Wilson *et al.* 1976, Blin *et al.* 1979
	Lytechinus variegatus	260	–	–	Patterson & Stafford 1970
Genes for 5s rRNA	*Lytechinus variegatus*	175	1.2-1.3 kb	0.025 %*	Nijhawan & Marzluff 1979
Genes for 5.8s rRNA	*Arbacia punctulata*	–	–	0.003-0.004 %	Sy & McCarty 1970
Genes for the histones	*Strongylocentrotus purpuratus*	1000/sperm	–	–	Farguhar & McCarthy 1973
	Strongylocentrotus purpuratus	400-500	6.1 kb	0.3 %	Kedes *et al.* 1975
	Strongylocentrotus purpuratus	–	approx. 6 kb	–	Weinberg *et al.* 1975, Overton & Weinberg 1978
	Strongylocentrotus purpuratus	–	6.54 kb	–	Cohn *et al.* 1976
	Echinus esculentus	–	7.1 kb	–	After Hentschel & Birnstiel 1981
	Psammechinus miliaris	1200	–	–	Weinberg *et al.* 1972
	Psammechinus miliaris	–	5.6 kb	–	Portman *et al.* 1976
	Psammechinus miliaris	–	6 kb	–	Schaffner *et al.* 1976, Gross *et al.* 1976
	Psammechinus miliaris	300-600	6.3 kb	–	After Hentschel & Birnstiel 1981
	Psammechinus (Paracentrotus) lividus	400	7.0 kb	–	After Hentschel & Birnstiel 1981
	Lytechinus pictus	–	7.2 kb	–	Cohen & Kedes 1979a, 1979b
	Lytechinus pictus	–	7.3 kb	–	Kedes 1979
Genes for actin	*Strongylocentrotus purpuratus*	5-20	–	–	Durica *et al.* 1980
	Strongylocentrotus purpuratus	⩾ 4	–	–	Schuler & Keller 1981
	Strongylocentrotus purpuratus	⩾11	–	–	Scheller *et al.* 1981a
Genes for α- and β-tubulins	*Lytechinus pictus*	9-13 for each	–	–	Alexandraki & Ruderman 1981

* These fractions of the genome were calculated from the repeat size, the reiteration frequency, and the published genome size (Hinegardner 1974).

\+ Kb is an abbreviation for kilobases or thousands of base pairs.

cribed extensively (Davidson *et al.* 1973a, Manning *et al.* 1975, Crain *et al.* 1976). Although the relative proportions of clustered (or long repetitive) and short repetitive DNA vary from species to species, the genomes of echinoderms, including echinoids, holothuroids, asteroids and ophiuroids, contain both long and short repetitive sequences and therefore possess the *Xenopus* pattern for sequence organization (Table 1) (Britten 1972, Davidson *et al.* 1973b, Vorob'ev & Kosjuk 1974, Davidson *et al.* 1975a, Lee *et al.* 1977, Brykov *et al.* 1978, Smith & Boal 1978, Smith *et al.* 1980).

2.2 *Repetitive sequence functions in echinoderms*

Clustered repetitive DNA comprises roughly 5-15 % of the genomes of echinoids, asteroids, and holothuroids (see Table 1). The genes of 26s, 18s, 5.8s, and the 5s ribosomal RNAs, the histones, and possibly for the actins and tubulins, should be included among the clustered repetitive DNA sequences of echinoderm genomes (see Table 2). However, the functions for the majority of the clustered repetitive sequences remain unknown. It is likely that clustered simple sequences (not present in significant quantities in the genomes of echinoderms) have little functional significance and may, therefore, be of minor importance for the phenotypic and evolutionary changes in response to natural selection (Miklos & Gill 1981). Furthermore the functions are not known for the short repetitive DNAs that are interspersed among the unique sequences and which comprise up to 25 % of the genomes of many higher eucaryotes. Some have suggested that these repetitive sequences may play an important role in the regulation of gene expression (Britten & Davidson 1968, Georgiev 1969, Robertson & Dickson 1975, Davidson & Britten 1979, Zuckerkandl 1981). Others have suggested that many of the repetitive sequences may be 'selfish' or 'parasitic' DNA (Orgel & Crick 1980, Doolittle & Sapienza 1980, Ohta 1981).

In brief, a large fraction of the genomes of most higher eucaryotes and echinoderms is composed of clusters of repetitive sequences plus short repetitive sequences interspersed among the unique DNA. Although mechanisms for the dispersion of repetitive sequences have been postulated (Jagadeeswaran *et al.* 1981, Pearson & Morrow 1981), the functions for the majority of these repetitive sequences remain unknown and thus, their evolutionary significance remains a matter for speculation.

3 GENETIC VARIABILITY IN UNIQUE DNA

3.1 *General considerations*

Unique DNA sequences, comprising 45-80 % of the genomes of many echinoderms (Table 1), include the majority of the structural genes that are trans-scribed and represented by messenger RNA in the polyribosome fraction of cellular homogenates (Goldberg *et al.* 1973, Davidson *et al.* 1975b). However, coding sequences probably comprise only a minor fraction of the total unique DNA (Galau *et al.* 1976b). Studies with the DNA of echinoderms and other eucaryotes indicate that a substantial amount of DNA may be located between fragments of the sequence coding for single proteins (Scheller *et al.* 1981a). These noncoding sequences are commonly referred to as 'introns' or 'intervening' sequences (for a review, see Lewin 1980). Additionally, there are unique sequences adjacent to or flanking both ends of the coding sequences. These flanking sequences likely contain short elements, less than 25 base pairs long, that act as promotors and polymerase binding sites (Busslinger *et al.* 1979, 1980). These short elements probably occur in the flanking zones adjacent to other structural genes and would thus technically be repetitive. Furthermore, the flanking regions contain the short, 250-350 base pairs long, interspersed repetitive sequences (Davidson *et al.* 1975b, Scheller *et al.* 1981a).

With respect to evolutionary changes in unique DNA sequences within echinoderms, coding sequences might be expected to diverge at different rates than the noncoding intervening and flanking sequences. Also, the measured rates for nucleotide substitions within coding sequences may vary due to differing numbers of potential nucleotide substition sites that would be classed as 'silent' vs. others classed as amino acid 'replacement' sites (Efstratiadis *et al.* 1980, Perler *et al.* 1980).

3.2 *Intraspecific variability in unique DNA*

A detailed analysis of unique sequence variability within a species of the Echinodermata has been reported by Britten *et al.* (1978) for *Strongylocentrotus purpuratus*. Their thermal denaturation studies of the reassociated unique sequence DNA of a single individual and of the cross-reacted unique sequences of two individuals, have established: 1) that the unique sequence variability (a form of genetic heterozygosity) in the genomes from diploid individuals are consistent with that expected for a large population with no evidence of inbreeding; 2) that the extent of interindividual sequence divergence involves nucleotide substitutions in about 4% of the nucleotide pairs; and 3) that the difference between two individuals from a single Southern Californian population is not measurably smaller than the difference between individuals from two distinct populations, one from Southern California and the other from near Victoria Island, British Columbia. The measured level of sequence variability between individuals (4%) was more than an order of magnitude higher than that calculated for the coding regions in genes with the largest levels of protein polymorphism (Grula *et al.* 1982). Because most larvae of echinoids are pelagic, there may be extensive gene flow from the populations of *Strongy-*

locentrotus purpuratus in British Columbia to the population in Southern California. Investigations utilizing individuals from populations experiencing little or no gene flow may yield empirical results different from those of Britten *et al.* (1978). Similar studies by Grula *et al.* (1982) demonstrate significant levels of unique sequence divergence between Atlantic and Pacific populations of *Strongylocentrotus droebachiensis.* However, the measurements by Grula *et al.* (1982) of the levels of DNA sequence variability, or 'polymorphism' occurring in *Strongylocentrotus droebachiensis* (2 %), *Strongylocentrotus franciscanus* (3.2 %) and *Strongylocentrotus intermedius* (3 %) are in agreement with the results reported earlier by Britten *et al.* (1978). The high levels of sequence variability measured presumably occur in the noncoding fraction of the unique sequence DNA. This conclusion is supported by results showing that the less variable unique sequences in the genome of *Strongylocentrotus purpuratus* exhibit lower levels of evolutionary sequence divergence between *Strongylocentrotus purpuratus* and *Strongylocentrotus franciscanus* than do the more variable sequences (Grula *et al.* 1982).

3.3 *Interspecific variability in unique DNA*

The rate of single copy sequence divergence in evolutionary lines leading to *Strongylocentrotus purpuratus* has been approximately 0.06-0.35 % per million years (Angerer *et al.* 1976) or 0.22 % per million years (Harpold & Craig 1978). Both studies reported that a small fraction of the unique sequences of *Strongylocentrotus purpuratus* has been conserved in a distant relative, *Lytechinus pictus* (12-14 % of the unique sequences cross-reacting and possessing 7.5-8.5 % mismatched base pairs).

Hall *et al.* (1980) investigated the unique sequence divergence between the genomes of *Strongylocentrotus purpuratus, S.droebachiensis, S.franciscanus* and *Lytechinus pictus* using procedures that minimize variables such as differences due to base composition. Their substantially lower criterion for reassociation (more than 40° below the melting point for native DNA) enabled recovery of a significantly larger fraction of the unique sequences in cross-species molecular hybridizations (i.e. 90 % of the tracer DNA of *Strongylocentrotus purpuratus* was recovered in hybrids with driver DNA of *Strongylocentrotus franciscanus* in contrast to 51-53 % recovered by Angerer *et al.* 1976, and Harpold & Craig 1978, for the same cross-reaction performed only 25-31° below the melting point of native DNA). Thus, Hall *et al.* (1980) included 40-45 % more of the unique sequences in their estimates for the rate of sequence divergence. As a consequence, their estimate of 0.5 % per million years differs from the earlier studies which considered only the more conserved sequences. However, the results are in general agreement that the rates for divergence of unique sequences are probably non-linear due to the complex nature of unique DNA.

Brykov *et al.* (1979b) examined the extent of unique sequence conservation among echinoids and asteroids. They reported that 53 % of the unique sequences of *Strongylocentrotus intermedius* are conserved in *Strongylocentrotus nudus* (with 10 % base pair mismatch), and 12 % are conserved in *Scaphechinus mirabilis* (an irregular echinoid) and *Asterias amurensis* (an asteroid). Smith *et al.* (1982) found that the extent of unique sequence DNA conservation of asteroids closely follows established taxonomic lines. They found that 11.8-12.1 % of the unique DNA sequences of *Pisaster ochraceus* are conserved in the distant relatives, *Solaster stimpsoni* and *Dermasterias imbricata.* All of the reports cited in this section agree that 11.8-14 % of the unique DNA has been conserved in even the most widely diverged species of echinoderms.

Angerer *et al.* (1976) reported that the unique sequences, complementary to gastrula polyribosomal messenger RNA, diverge, at a lower rate than the total unique sequences, between closely related species pairs. However, the sequences conserved in distantly related species are not enriched in their content of structural gene sequences. This latter conclusion confuses, rather than clarifies, the logical deduction by Grula *et al.* (1982) that the more slowly evolving unique sequences probably have sequence-dependent functions (i.e. sequences coding for proteins).

4 VARIABILITY IN REPETITIVE DNA SEQUENCES

4.1 *Potential errors in interpreting DNA-DNA reassociations*

Britten & Kohne (1968) reported the existence of intragenomic divergence among the repeated sequences of eucaryotes. The quantity of repetitive sequences present within a genome that is detected by traditional molecular reassociation kinetic techniques is affected by the 'criterion' for reassociation (salt concentration, melting point for the native DNA and the reassociation temperature are implicit in the criteria for reassociation). Also, the amount of intragenomic divergence (measured as base-pair mismatch) in the repetitive sequences will be influenced by the criterion for reassociation. Thus, one must be cautious about interpreting estimates for base-pair mismatch within repetitive sequence families. Additional considerations must include the length of DNA and the arrangement of sequences within large fragments of DNA. These can affect the stability of duplexes formed during preparative reassociations (Chaudhari & Craig 1979b, Moyzis *et al.* 1981). Failure to consider these factors has probably contributed to oversimplification of the organization of clustered repetitive sequences and their evolution. Specifically, after initial purification, long repetitive sequences have a thermal stability close to that for native DNA (Britten & Davidson 1976, Britten *et al.* 1976, Eden *et al.* 1977). This was interpreted to result from sequence homogeneity among the members of the repetitive sequence families. This conclusion led to the seemingly logical assumption that many long repetitive DNA sequences must have arisen by recent amplification events within the genomes of the echinoids studied.

4.2 *The organization and evolution of long repetitive DNA*

The conclusion that long repetitive DNA of echinoids had arisen by recent amplification events witihn the genomes was disputed (Chaudhari & Craig 1979a, Craig *et al.* 1979, Chaudhari & Craig 1979b), and has been refuted (Scheller *et al.* 1981b). Scheller *et al.* (1981b) used cloned repetitive sequences from *Strongylocentrotus purpuratus* to prove that some of the long repetitive sequences are composed of assemblages of smaller (short) repetitive sequences. Although these short repetitive sequences may appear adjacent to other repetitive or unique sequences at other locations in the genome, they appear in the same contiguous colinear order (with respect to the other short repetitive elements; see Fig.1). These colinear arrays of short repetitive sequences may be several kilobases in length, and are referred to as 'subfamilies' of repetitive DNA. Analysis of several subfamilies of colinear long repetitive sequences indicate that most probably contain 25 ± 10 members in the genome (Scheller *et al.* 1981b).

The existence of such colinear arrays of short repetitive sequences within the 'subfamilies' of long repetitive DNA was predicted by Craig *et al.* (1979) and, as demonstrated by Chaudhari & Craig (1979b), these arrays were probably responsible for the formation of stable duplexes by branch migration during reassociations of long fragments of DNA. 'Given long, precisely arrayed sets of repetitive sequences, including sequences which may form duplexes with members of the more highly reiterated short repetitive sequences, there would still remain a finite probability of nucleations forming between exact complements within a long repetitive sequence (see segment d in Fig.4-II). Once nucleated, the DNA would zipper, or branch migrate (Broker 1973), from end to end, unless interrupted by sequences which were out of order in the total array or by sequences which were less redundant (e.g. unique sequences) . . . Nevertheless, the zippering of perfectly matched duplexes should proceed with the exclusion of any less stable, mismatched sequences' (Craig *et al.* 1979).

Scheller *et al.* (1981b) demonstrated that one of the colinear long repetitive sequences from the genome of *Strongylocentrotus purpuratus* has been precisely maintained in the genome of *Strongylocentrotus franciscanus* (Fig.2). The genome of *Tripneustes gratilla* may possess the same colinear long repeat found in *Strongylocentrotus purpuratus* although it contains more less-closely related sequences. These results demonstrate that this colinear long repeat is evolutionarily ancient and is shared not only among congeneric species of echinoids but even among distantly related species belonging to diverse taxonomic families (Scheller *et al.* 1981b). The general applicability of the above

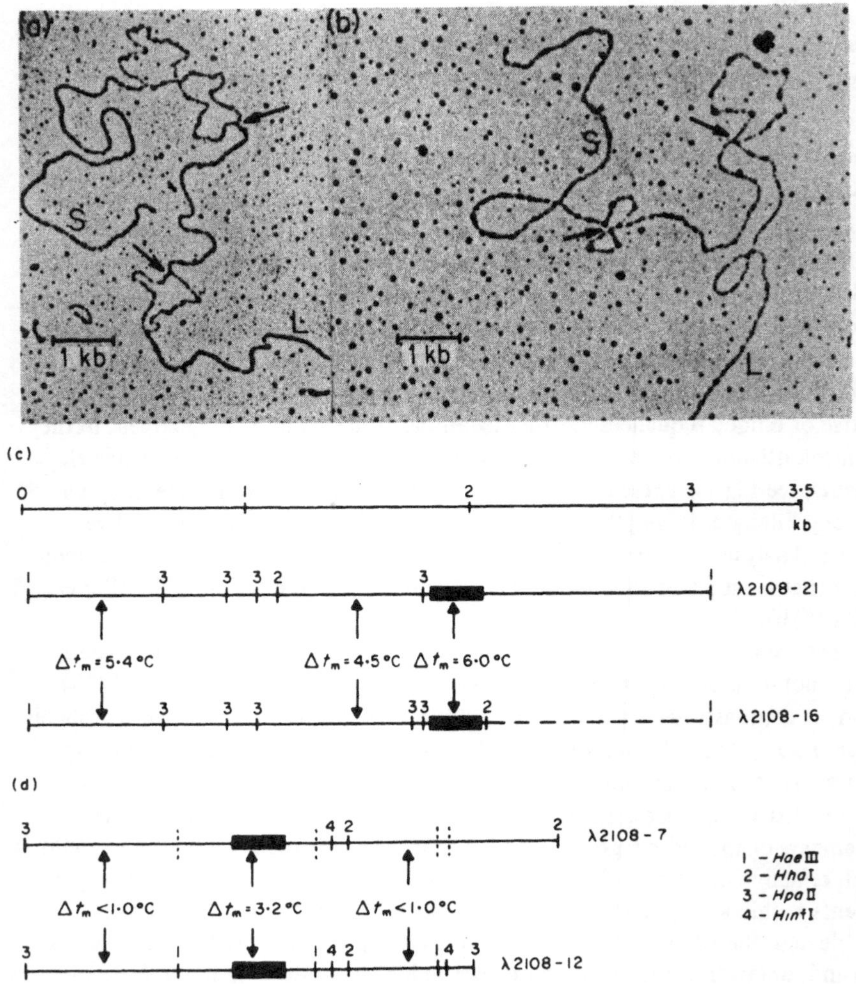

Figure 1. Colinearity and sequence homology among members of cloned repeat subfamilies of DNA from *Strongylocentrotus purpuratus* (designated λ2108-xx subfamilies). a) Heteroduplex formed between λ2108-32 and λ2108-33. Arrows indicate the limits of the colinear repetitive sequence. L: long arm (left arm) of λ phage vector; S: short arm of vector. The repetitive sequence is terminated on both ends by non-homologous (single-stranded) echinoid DNA. b) Heteroduplex formed between λ2108-12 and λ2108-7. The colinear repetitive sequence and vector arms are indicated as in (a). The repeat element of λ2108-7 is terminated on one end by the long arm of the vector, while that in λ2108-12 is terminated at the other end by the short arm of the vector. Therefore, the loops at the ends of repeat duplex consist in each case of echinoid DNA from only one of the recombinants rather than from both as in (a). c) Comparison of restriction sites and heteroduplex thermal stabilities within a 3.2 kb (kilobase) *Hae*III restriction digested fragment of the colinear repeats in λ2108-16 and λ2108-21. The broken line indicates a region of the λ2108-16 fragment where restriction sites were not mapped. ΔT_m is the difference between the ther-

conclusions relating to the evolutionary conservation of colinear long repetitive sequences was reported by Chaudhari (1980). Chaudhari concluded that some of the precise arrays of repetitive sequences within the long repetitive sequences of *Lytechinus variegatus* have been maintained within the genomes of *Lytechinus pictus, Sphaerechinus granularis, Paracentrotus lividus, Psammechinus microtuberculatus,* and *Arbacia lixula.* However, she reported that some of the arrays of shorter elements within long repeats may have been rearranged during evolution (Fig.3, and Tables 3 and 4). Variable arrangements of arrays of clusters of short repetitive sequences have also been reported in the genomes of fruit flies (Wensink *et al.* 1979) and birds (Musti *et al.* 1981, Sobieski & Eden 1981).

These results, particularly those of Scheller *et al.* (1981b), indicate that the model proposed by Craig *et al.* (1979) for the organization of shorter elements within the long repetitive DNA of echinoids is fundamentally correct (Fig.4).

4.3 *The relationship between long and short repetitive DNA*

The differences in thermal stability between the long and short repetitive DNA sequences of echinoderms have been reported by a number of investigators (Britten *et al.* 1976, Eden *et al.* 1977, Brykov *et al.* 1978, Smith & Boal 1978, Chaudhari & Craig 1979a, b, Craig *et al.* 1979, Brykov *et al.* 1979b, Poltaraus *et al.* 1980a, b, Smith *et al.* 1980, Moore *et al.* 1981). Earlier reports interpreted this difference in stability to support the hypothesis that long and short repetitive sequences are principally independent classes of DNA. This hypothesis was put forth despite results of molecular hybridization analyses demonstrating that 10% of the long repetitive sequences occurred in the short repetitive DNA of *Strongylocentrotus purpuratus* (Eden *et al.* 1977). Eden *et al.* (1977) assumed that this represented cross contamination in the short repetitive DNA. Reverse reactions, using short repetitive DNA tracer against long repetitive driver to determine the percent of the short repetitive sequences in the long repetitive DNA, were not reported by Eden *et al.* (1977).

---

mal stability of the renatured 5′ terminally labelled λ2108-16 probe sequences and the thermal stability of the heteroduplex formed when these fragments were reacted with the 3.2 kb *Hae*III fragment of λ2108-21. The probe fragments are indicated by arrows and these fragments were shown to react only with the corresponding regions of the λ2108-21 sequence in an independent gel blot experiment. The solid box indicates the position of the sequence subelement which reacts with the CSp2108 insert d) Comparison of restriction sites and heteroduplex thermal stabilities in a region of the colinear repetitive sequences in λ2108-7 and λ2108-12. Notation and experimental procedures are as in (c). Broken lines indicate sites that were not mapped in the λ2108-7 fragment and may or may not be present in this fragment. The location of the sequence reacting with the CSp2108 probe is not known exactly except that it is between the two *Hae*III sites indicated in the λ2108-12 fragment (from Scheller *et al.* 1981b).

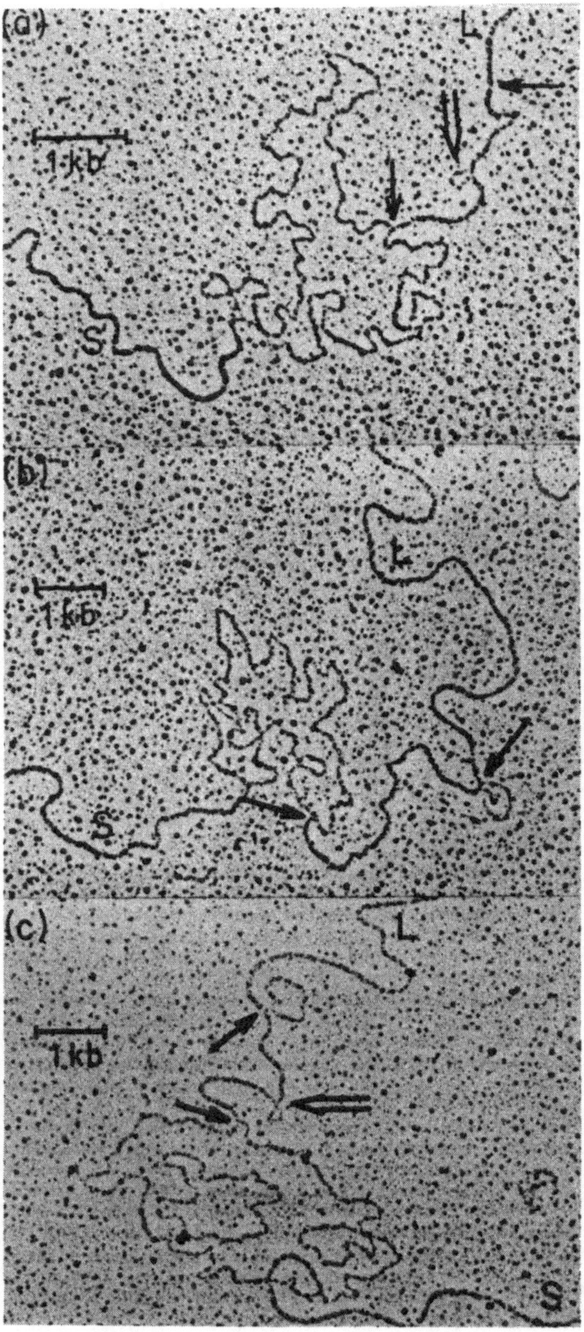
(a)
L
1 kb
S
(b)
L
1 kb
S
(c)
L
1 kb
S

Studies of cloned repetitive sequences demonstrated that some short repetitive sequences are embedded in longer repetitive DNA and also occur singly, interspersed in unique DNA (Anderson *et al.* 1981). Although some sequences within long and short repetitive DNA may be exclusive to those fractions, some of the repetitive elements can be found in both types of arrays (Anderson *et al.* 1981, Chaudhari 1980). Other than the disagreement with respect to the existence of sequence overlap between long and short repetitive DNA, there is little disagreement about the organization and evolution of the short interspersed repetitive sequences within the genomes of echinoderms.

4.4 *The organization and evolution of short repetitive DNA*

The low thermal stability of the short repetitive DNA, after preparative reassociation, indicates that the copies of these repetitive sequences in echinoderm genomes are internally diverged (Britten & Kohne 1968, Britten & Davidson 1976). Britten & Davidson (1976) proposed that these diverged copies resulted from a history of base substitution and translocation. A study of intragenomic divergence of individual repetitive sequence familes in the genome of *Strongylocentrotus purpuratus* showed that of 18 cloned repeats, three belonged to families internally diverged by less than 4% and 13 belonged to families that were internally diverged by less than 10% (Klein *et al.* 1978). The 13 cloned repeats seemed to lack any detectable highly divergent sequence relatives in the genome. The remaining five of the 18 clones belonged to families of widely diverged repetitive sequences. These results suggest that families of repetitive sequences might result from the saltatory expansion of ancestral sequences at

---

Figure 2. Heteroduplexes with cloned DNA from the genome of *Strongylocentrotus purpuratus* recombined with DNA from λ-phage. a) One recombinant (λSP2108-21R which contains the complete 5.2 kilobase element of the *Strongylocentrotus purpuratus* CSp subfamily) was reacted with a cloned fragment (λSf2108-13) from *Strongylocentrotus franciscanus,* which according to other studies contains a sequence closely related to the CSp subfamily. The 5.2 kilobase repeat sequence of λSf2108-13 is contiguous with the left arm of the λ vector. The heavy arrow mark on the left end the approximate junction between vector and insert DNA, and on the right the termination of the region of homology. The open arrow indicates a 580 nucleotide insert in the *Strongylocentrotus franciscanus* sequence. b) Heteroduplex of λSp2108-21R with λSf2108-11. Notation as in (a). This example of the *Strongylocentrotus franciscanus* sequence lacks the insert sequence observed in (a). The junction between the 5.2 kilobase repeat element and the long arm of the λ vector is now visible as a loop formed at the terminus of the repeat, which in this *Strongylocentrotus franciscanus* insert occurs several hundred nucleotides prior to the terminus of the insert. c) Heteroduplex of λSf2108-11 and λSf2108-13. The end of the 5.2 kilobase region of homology near the long arm of the λ vector is again marked by a loop of echinoids DNA due to the different termini in the two *Strongylocentrotus franciscanus* inserts, and the 580 nucleotide insertion is marked as in (a) (from Scheller *et al.* 1981b).

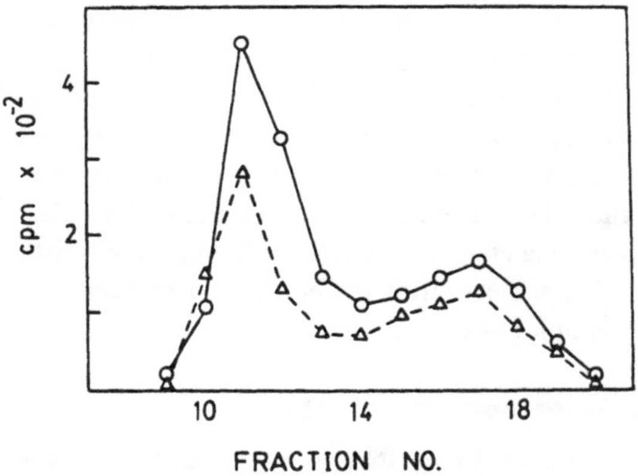

Figure 3. Elution profile of homologous and heterologous duplexes from Biogel A-50. (^{3}H)-labelled DNA from *Lytechinus variegatus* was sheared to approximately 5000 nucleotide lengths and was reassociated to Cot 10 with a 500 fold excess of either *Lytechinus variegatus* (o) or *Lytechinus pictus* (Δ) long native DNA in NaCl-Pipes buffer at 65°. These were treated with S_1 nuclease to a digestion estimate, or DIG, of 0.95 (Craig *et al.* 1979), and the double stranded hybrids were bound to hydroxylapatite and eluted with 0.41M Phosphate buffer. These duplexes were subsequently chromatographed on a column of Biogel A-50 and the radioactivity in 250 μl aliquots of the 5 ml fractions was determined. The fraction of radioactivity eluting in the region expected for long repeated DNA was calculated and appears in Table 3 (from Chaudhari 1980).

different times in the evolutionary history of the species (Klein *et al.* 1978). Although 18 clones may be inadequate for statistical significance, the grouping of the repetitive sequence families into three levels of sequence divergence may be indicative of periodic saltatory amplification of sequences during the evolution of *Strongylocentrotus purpuratus.*

Studies of interspecies repetitive sequence divergence in echinoderms indicate that the repetitive sequences have been more conservative than the total unique sequences during the evolution of species (Harpold & Craig 1977, Moore *et al.* 1978, Chaudhari & Craig 1979a, Brykov *et al.* 1979a, b, Poltaraus *et al.* 1980a, b, Moore *et al.* 1981). However, there is some disagreement as to the fraction of repetitive sequences that are co-represented in the genomes of related species pairs. Differences in reassociation criteria, Cot values (products of the concentration of DNA in moles of nucleotides multiplied by the time in seconds at the termination of crossreactions), and differences in the ratios for the concentration of the radiolabelled 'tracer' DNA from one species relative to the concentration of unlabelled 'driver' DNA from the other, all complicate direct comparisons of the results. Possibly due to one or more of these factors, the results reported by Brykov *et al.* (1979a, b), for the high levels of repetitive sequence conservation in species from different classes of echinoderms, disagree with the majority of other reports. In the study by Brykov *et al.* (1979a), the

Table 3. Reassociation between long fragments of (^{3}H)-labelled DNA of *Lytechinus variegatus* and similarly sized driver DNA's of other species.

Driver species	% double stranded	% of duplexes longer than 1000 bases	% of input cpm found as long duplexes	% of tracer maintained as common arrays
None (control)	7.1	51	4.0	–
Lytechinus variegatus	25.1	48	12.0	8.0
Lytechinus pictus	16.0	51	8.0	4.0
Sphaerechinus granularis	10.3	56	5.8	1.8
Paracentrotus lividus	8.2	76	6.2	2.2
Psammechinus microtuberculatus	7.8	75	5.9	1.9
Arbacia lixula	7.4	75	5.6	1.6

The Cot at the termination of these reactions was too low to permit significant levels of reassociation of unique sequences, which should reassociate between Cot 100 and Cot 10,000 in these species.

A control reaction was set up in which the same quantity of tracer DNA was used, but no driver DNA was added. Instead, reassociation buffer was added to adjust the final concentration of tracer DNA to that found in all other reactions. In this control reaction, 7.1 % of the input radioactivity was duplexed. This reflects selfannealing of the tracer, caused by foldback sequences as well as by very high frequency repeats. The fraction of total input radioactivity that is found in long duplexes was calculated for each reaction as the product of % cpm double stranded and % duplexes that are larger than 1000 basepairs.

The chromatographic separations of the long and short duplexes formed, using *Lytechinus variegatus* and *Lytechinus pictus* 'driver' DNAs appear in Figure 3. It is clear that when *Lytechinus pictus* is the driver, in such a reaction, only 16 % of the input radioactivity is found in duplexes and of this, 51 % in duplexes larger than about 1000 nucleotides. Both these values are lower than the corresponding ones when homologous *(Lytechinus variegatus)* driver is used. Similar reactions were also set up using 5000 rpm sheared unlabelled total DNAs of several other species. From Table 3, it is evident that as phylogenetic distance increases between the tracer and driver species, the total amount of conserved arrays of clustered nonidentical repetitive sequences drop (from Chaudhari 1980).

significant increases in radiolabelled tracer DNA duplex formation at high Cot values might be due to self association of the tracer DNA. This possibility is supported by the remarkably high melting points for the supposed cross-species DNA molecular hybrids. In contrast, all other studies report relatively low melting points for the molecular hybrids between the repetitive sequences of widely diverged species pairs (Harpold & Craig 1977, Moore *et al.* 1978, Chaudhari & Craig 1979a, Poltaraus *et al.* 1980a, b, Moore *et al.* 1981). However, certain tandemly repeated genes, coding for rRNA and the histones, would be expected to be mutually conserved in widely diverged species pairs (see section 4.5). These sequences may account for much of the repetitive sequence homology that has been detected between widely diverged species pairs (Chaudhari & Craig 1979a, Poltaraus *et al.* 1980a).

Table 4. Melting temperatures of representative self- and cross-species DNA hybrids.

Tracer species	Driver species	T_m of long duplexes	T_m of short duplexes
Lytechinus variegatus	*Lytechinus variegatus*	85.0	78.0
Lytechinus variegatus	*Lytechinus pictus*	84.0	75.0
Lytechinus variegatus	*Sphaerechinus granularis*	82.5	71.5

In order to confirm that the long duplexes produced in the reactions appearing in Table 3, were true long repeat sequences, the melting temperatures (T_m) for some representative duplexes were determined after shearing them to approximately 350 basepairs. The results, in Table 4, show that using *Lytechinus variegatus* tracer and *Lytechinus pictus* driver, the short duplexes produced have a T_m of 75°. This is a drop in T_m of 3° below that of homologous short repeat DNA, and indicates that many of the duplexes assayed are cross species hybrids. On the other hand, the long duplexes produced during the cross-reaction have a T_m of 84° which is very close to that of the homologous long repeats. Four percent of the input radioactivity was calculated (8 % minus 4 % control value from Table 3) to be in long cross-species duplexes between the two *Lytechinus* species. This would represent about half of all duplexes formed. The other half might represent self-annealing of the tracer. In the case of *Sphaerechinus granularis* driver, the depression in T_m amounting to 2.5° for long duplexes and 6.5° for short ones, is even more pronounced. This is as expected with the greater evolutionary distance between the two species. The fact that the average melting temperature of the long duplexes produced is lowered supports the contention that a significant portion of the tracer must have formed hybrids with heterologous driver molecules, as self-reassociation would probably yield a T_m closer to that for the homologous *(Lytechinus variegatus)* reaction. This indicates that some arrays of repeating units are indeed found in common between *Lytechinus variegatus*, *Lytechinus pictus*, and *Sphaerechinus granularis* (from Chaudhari 1980).

The majority of long and short repetitive DNAs diverge at nearly equivalent rates between echinoderm species (Chaudhari & Craig 1979a). This conclusion is supported by results showing that long repetitive sequences have as great an interspecies sequence difference as the less precise short repetitive sequences (Moore *et al.* 1981).

Some of the genomic variability between species of echinoderms is a consequence of very significant differences in the repetition frequency of sequences belonging to specific families of sequences in different species (Moore *et al.* 1978, Moore *et al.* 1981). Thus, repetitive sequences prominent in one species are often present in much lower frequency in a closely related species. Moore *et al.* (1981) concluded that many of the large families of repetitive sequences in *Strongylocentrotus purpuratus* are those which have grown in size since the divergence of *Strongylocentrotus purpuratus* and *Strongylocentrotus franciscanus.* Whether these changes in repetition frequency have any evolutionary significance is not clear because these species have comparable levels of specific repetitive sequences represented in their egg RNA's despite the differences in numbers of copies of the repetitive sequences in the genomes (Moore *et al.* 1980).

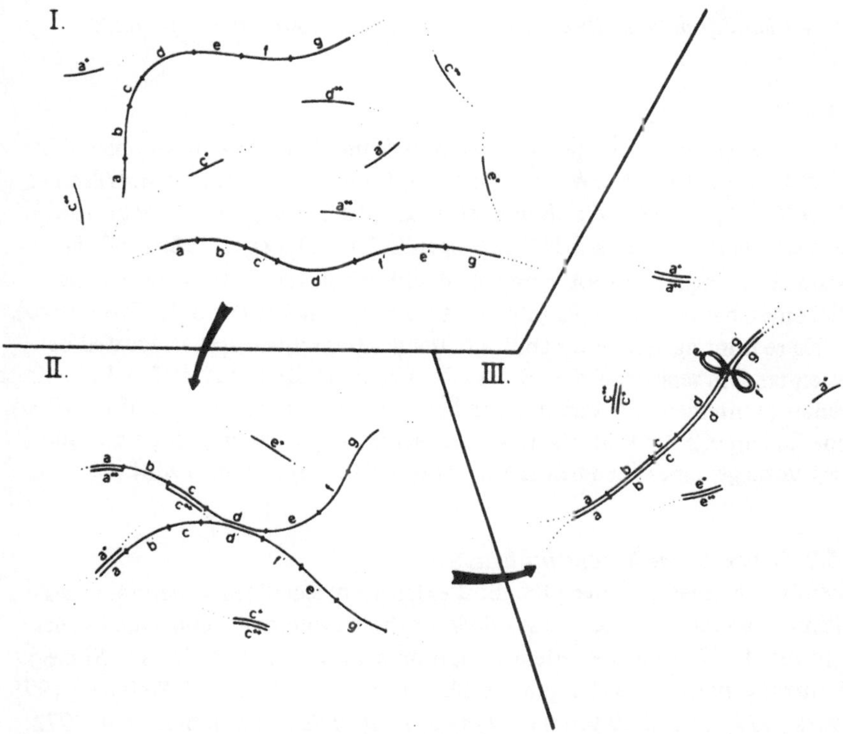

Figure 4. A model for the sequence organization within long repetitive DNA. The solid lines represent single strands of repetitive DNA. Contiguous unique sequences are represented by a dotted line. Segments of repetitive sequences which represent perfect complements of one another are designated by letters, either with or without primes. Imperfect complements, belonging to the same repetitive sequence families, are represented by letters with single and double asterisks. Roman numerals I, II and III denote increasing times of reassociation going from low to high Cot values. Due to zippering, or branch migration, started at sequence d and d′ (in II), short mismatched repetitive sequences formed previously are displaced, leaving long perfectly matched, as well as short mismatched, double stranded DNA (in III) (from Craig *et al.* 1979).

Although several models have been proposed for the function of repetitive sequences (Britten & Davidson 1968, Georgiev 1969, Davidson & Britten 1979, Zuckerkandl 1981), proof of function is lacking for most of the repetitive sequences and, therefore, it is impractical to speculate about the functional or evolutionary significance of 1) sequence conservation in repetitive DNA, 2) evolutionary changes in repetition frequency among members of specific repetitive sequence families, 3) structural relationships between long and short repetitive DNA, or about 4) the existence of highly conserved colinear arrays of short repetitive sequences within long repetitive DNA.

4.5 *Variability in 'identified' sequences belonging to multigene families*

4.5.1 *Genes for rRNA*

Genes coding for the 26 and 18s rRNAs in echinoderms have been isolated by various means (Patterson & Stafford 1970, Joseph & Stafford 1976, Wilson *et al.* 1976, Edgell *et al.* 1978, Blin *et al.* 1979, Mishra 1979, Bieber *et al.* 1981). Also the genes for the 5s rRNA have been isolated (Lu *et al.* 1980) and the existence of the 5.8s rRNA, complexed with the larger rRNA in eucaryotic ribosomes, has been demonstrated in *Arbacia punctulata* (Sy & McCarty 1970).

There is intragenomic variability in the nontranscribed spacer, located between tandem repeats of the 18-26s rRNA genes (Edgell *et al.* 1978). In contrast studies of intergenomic variability indicate that the coding regions of the rRNA genes are among the most rigorously conserved sequences in procaryotic and eucaryotic genomes (Bendich & McCarthy 1970, Petrov *et al.* 1980).

4.5.2 *Genes for the histone proteins*

The histone genes have been the most extensively examined structural genes within echinoderms. The genes coding for these proteins are clustered (Kedes & Birnstiel 1971) and are reiterated approximately 300-1000 times in *Strongylocentrotus purpuratus* (Farguhar & McCarthy 1973; Overton & Weinberg 1978, Kedes 1979), 300-1200 times in *Psammechinus miliaris* (Weinberg *et al.* 1972, Hentschel & Birnstiel 1981), and 400 times in *Psammechinus (Paracentrotus) lividus* (Hentschel & Birnstiel 1981). Hybridizations of histone messenger RNA from *Strongylocentrotus purpuratus* with the DNA of *Strongylocentrotus droebachiensis* and *Lytechinus pictus* (Farguhar & McCarthy 1973) and hybridizations of the histone mRNA of *Psammechinus miliaris* with the DNA of *Paracentrotus lividus* (Weinberg *et al.* 1972) have established that there has been little drift in the sequences coding for these proteins during the evolution of these species. Farguhar & McCarthy (1973) stated that the low levels of third position changes suggest that selection pressure may act at the level of the nucleic acid sequence in these genes.

The fine structural details of cloned histone genes have been described (Kedes *et al.* 1975, Weinberg *et al.* 1975, Cohn *et al.* 1976, Portman *et al.* 1976, Schaffner *et al.* 1976, Gross *et al.* 1976, Schaffner *et al.* 1978, Cohn & Kedes 1979a, Mellado *et al.* 1981). These studies, utilizing the DNA from *Strongylocentrotus purpuratus, Psammechinus miliaris, Lytechinus pictus* or *Echinus esculentus,* demonstrated that all five histone genes are contained within a 5.6-7.2 kilobase fragment of DNA. This fragment appears 400-500 times in the genome of *Strongylocentrotus purpuratus* (Kedes *et al.* 1975). The order of the histone genes is H_{2A}-H_3-H_{2B}-H_4-H_1 in the DNA of *Psammechinus miliaris* (Gross *et al.* 1976), *Strongylocentrotus purpuratus* (Cohn *et al.* 1976), *Lytechi-*

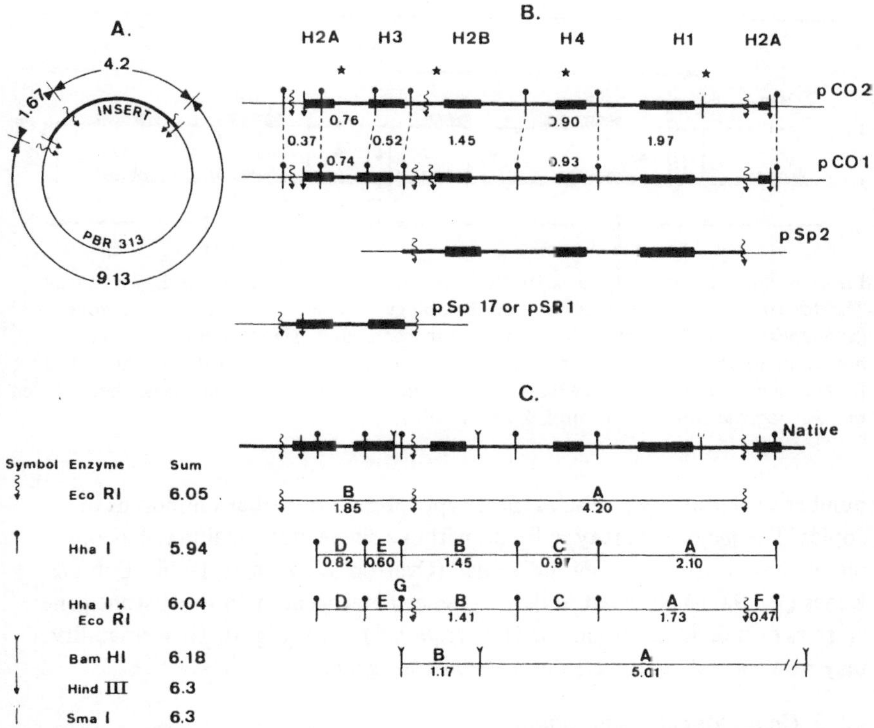

Figure 5. Restriction enzyme analysis of cloned and native histone DNA from the echinoid, *Strongylocentrotus purpuratus.* A heavy solid line represents the histone gene inserts; the lighter line represents vector DNA; and the solid blocks indicate the position of the mRNA sequences as determined by electron microscopic studies of hybrids. A) The orientation of the histone gene insert with respect to the vector, pBR313. Only Eco RI (ξ) and Hind III (↓) sites are shown. B) Restriction enzyme map of recombinant clones designated pCO1 and pCO2. The regions cloned previously and designated as pSp2, pSp17 and pSR1 are also indicated. Values between the dashed lines refer to the size of Hha I restriction fragments, in kilobases, of pCO1 and pCO2. Asterisks indicate four positions of difference in the two plasmids. C) Restriction enzyme map of enriched DNA (from Overton & Weinberg 1978).

nus pictus (Cohn & Kedes 1979a), and *Echinus esculentus* (Mellado *et al.* 1981). Conserved regulatory sequences probably exist near the 3′ end (Busslinger *et al.* 1979) and the 5′ flanking sequence adjacent to the histone genes (Busslinger *et al.* 1980). No intervening sequences occur within the histone genes of *Lytechinus pictus* (Cohn & Kedes 1979b). Overton & Weinberg (1978) detected intra-, as well as intergenomic heterogeneity in the histone gene repeats of *Strongylocentrotus purpuratus* (Fig.5). There are small differences in repeat lengths due to changes in the spacer regions between genes, and there is no repeat class common to all five individuals tested. Each individual has ‘a small

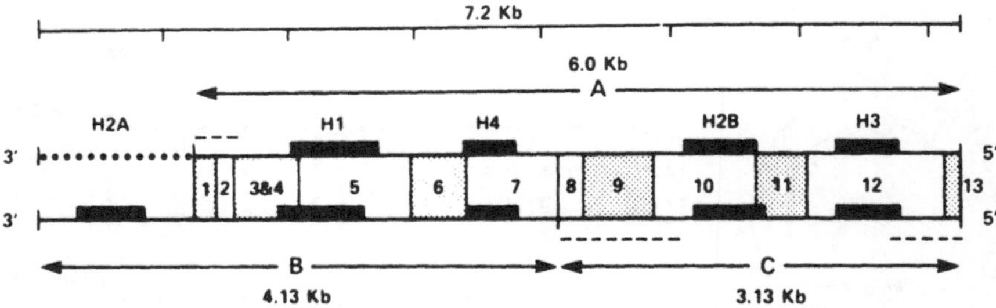

Figure 6. Summary map of the variant histone genes of *Lytechinus pictus* drawn to scale. The dotted areas represent regions of nonhomology between the two types of histone gene repeat units. The dashed lines represent spacer areas which show little sequence homology to *Strongylocentrotus purpuratus* histone genes. The heavy dotted line is Eco R1 fragment (Cohn & Kedes 1979a), which contains the H2A gene and has not been cloned into *Escherichia coli* (from Cohn & Kedes 1979b).

number of repeat types', each repeat type 'present in perhaps hundreds of copies. The major repeat type found within a single individual may be completely absent from another individual' (Overton & Weinberg 1978). Cohn & Kedes (1979a, b) observed similar sequence heterogeneity in the histone gene clusters of individual specimens of *Lytechinus pictus* (Fig.6). This variability may be a general characteristic of the histone genes.

4.5.3 *Genes for actin and tubulin*

The genes for actin and tubulin are not as highly reiterated as those for rRNA or the histones. However, their coding sequences have been conserved adequately to allow molecular hybridization with cDNA probes prepared from embryonic chicken brain actin or tubulin mRNAs (Cleveland *et al.* 1980).

Hybridization analyses, using cloned α- and β-tubulin genes indicate that there are at least 9-13 complete sequences for each of the two tubulin gene families per haploid genome of *Lytechinus pictus* (Alexandraki & Ruderman 1981). Although the α- and β-tubulin genes do not appear to be closely linked, some members within the α- and β-tubulin gene families are clustered. Molecular hybridizations performed at different criteria suggest the existence of heterogeneous, ancestrally related, α- and β-tubulin genes within the genome of *Lytechinus pictus* (Alexandraki & Ruderman 1981).

Durica *et al.* (1980) estimated that there are 5-20 actin genes per haploid genome in *Strongylocentrotus purpuratus.* Schuler & Keller (1981) determined that at least two of these genes are linked and possess intervening sequences beginning after the codon for the 121st amino acid in the actin sequence. Scheller *et al.* (1981a) analyzed 11 different actin genes from the DNA of *Strongylocentrotus purpuratus* (Fig.7). Although the coding sequences of all 11 clones are very similar, variability in the intervening and flanking sequences demonstrate that there are at least five distinct types of actin genes.

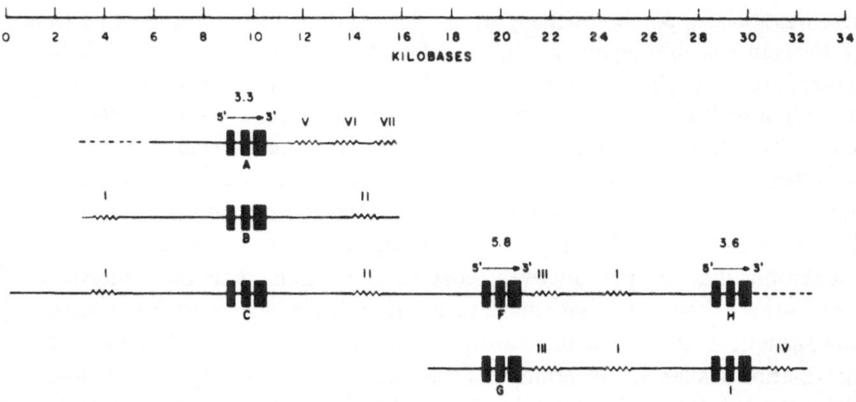

Figure 7. Repetitive sequences flanking actin genes A to F. Repeat sequence elements of families I to VII are represented by jagged lines, and the actin gene protein coding region is indicated by solid blocks, as in earlier figures. It is assumed that the 32 kilobase region containing genes C, F and H is actually a contiguous occurrence in the genome, which is not unequivocally shown, but is probable. However, no logical consequence of significance for this analysis of the pattern of repeat sequence occurrence would follow even if genes C and F were not linked directly to H, i.e. if H were linked instead to another copy (F′) of the 5.8 type gene F. The 5.8, 3.6 and 3.3 type genes refer to the number of kilobases between Eco R1 restriction sites flanking these genes. As indicated, genes A, B and C are of the 3.3 type, genes F and G are of the 5.8 type, and genes H and I are of the 3.6 type. Since the region occupied by repeat I is not included on the λSpG2-28 cloned fragment, a dashed line is shown in place of repeat I to the left of gene A (from Scheller *et al.* 1981a).

The position of intervening sequences within evolutionarily related genes is a generally conservative feature of gene evolution (Leder *et al.* 1980, Royal *et al.* 1979, Whali *et al.* 1980). Intervening sequences occur between codons 121/122 and 203/204 in most echinoid actin genes. The positions of these intervening sequences do not conform with their positions in actin genes of other organisms for which data are available (Scheller *et al.* 1981a, Davidson *et al.* 1982). 'This suggests that though the actin gene itself is very old, the evolutionary gene duplication process which gave rise to the *Strongylocentrotus purpuratus* gene set occurred since the divergence of the deuterostomes and perhaps even within the radiation of the echinoderms' (Scheller *et al.* 1981a).

5 CONCLUSIONS

The understanding of genomic variability and change is clearer and more concordant for echinoderms than for most other phyla. One would hope that we are on the verge of elucidating the mechanisms of genome function and evolution, and it is likely that echinoderms will continue to be a useful system for investigations supportive of these efforts. For example, correlations of patterns of divergence within repetitive sequence families, with fossil records of cladogenetic events, might lead to a better understanding of the molecular changes associated with species differentiation. Perhaps, patterns of sequence divergence within the genomes of extant species will provide molecular evidence that is consistent with the punctuations in fossil records that have been discussed by Eldredge & Gould (1972). Also the study of intra- and interindividual sequence divergence in the DNA from specimens belonging to large as well as small populations that are reproductively isolated may help increase our understanding of some of the molecular changes occurring that may be relevant to gene flow, genetic drift, and the differentiation of small isolated populations. The information available, concerning the evolutionary histories of extant echinoderms, in addition to the expanding body of knowledge about their genomic organization and its evolution, demonstrate that echinoderms are likely to be used extensively in such future studies.

6 APPENDIX

Many sequences, considered repetitive, are not identical copies of the same sequence but are descendants of a specific sequence which was amplified in the evolutionary past and, due to base pair substitutions, has diverged from the other copies or members of that 'family' of repetitive sequences. Eventually these base pair substitutions lead to sufficient divergence between the members of the 'repetitive sequence family' such that they can no longer be characterized as repeated using traditional methods of isolation. These methods involve the separation of complementary strands of DNA followed by reassociations under conditions or 'criteria' that permit the formation of hybrids between strands that may not be exact complements of one another. Hybrids formed from 'complements' that have diverged from one another will be less stable than perfect hybrids involving sequences that reassociate with their original complements. In practice, at 25° below the melting point of the native DNA ('standard criteria'), strands which fail to complement each other at fewer than 75-85 % of their bases will likely form only temporary hybrids. Thus, widely diverged copies of a particular sequence will not cross react and the fact that they belong to the same repetitive sequence family will not be detected. For this reason, reassociations of bovine DNA at 8° below the melting point of the native DNA identify less than 10 % of the genome as repetitive while reassociations at 36° below the melting point identify 55 % of the genome as repetitive (Britten & Davidson 1971).

In summary, depending on the criterion for reassociation, which is dependent upon the difference between the melting point and the temperature used for reassociation, large variations in the measured amount of repetitive sequences as well as in the calculated reiteration frequency and in the stability of duplexes formed are likely to be observed. However, it has been suggested that the widely diverged repetitive sequences, which are characterized by the investigator as unique sequence DNA, might also be recognized by the organism as unique sequence DNA (Bendich & Anderson 1977).

7 REFERENCES

Alexandraki, D. & J.V.Ruderman 1981. Sequence heterogeneity, multiplicity, and genomic organization of α- and β-tubulin genes in sea urchins. *Molec. cell. Biol.* 1:1125-1137.

Anderson, D.M., R.H.Scheller, J.W.Posakony, L.B.McAllister, S.G.Trabert, C.Beall, R.J.Britten, & E.H.Davidson 1981. Repetitive sequences of the sea urchin genome. Distribution of members of specific repetitive families. *J. mol. Biol.* 145:5-28.

Angerer, R.C., E.H.Davidson & R.J.Britten 1976. Single copy DNA and structural gene sequence relationships among four sea urchin species. *Chromosoma* 56:213-226.

Bendich, A.J. & R.S.Anderson 1977. Characterization of families of repeated DNA sequences from four vascular plants. *Biochemistry* 16:4655-4663.

Bendich, A.J. & B.J.McCarthy 1970. Ribosomal RNA homologies among distantly related organisms. *Proc. natl Acad. Sci. US* 65:349-356.

Bieber, D., N.Blin & D.W.Stafford 1981. The region of transcriptional initiation in *Lytechinus variegatus* rRNA genes. *Biochem. Biophys. Acta* 655:366-373.

Blin, N., J.M.Sperrazza, F.E.Wilson, D.G.Bieber, F.S.Mickel & D.W.Stafford 1979. Organization of the ribosomal RNA gene cluster in *Lytechinus variegatus.* Restriction analysis and cloning of restriction fragments. *J. biol. Chem.* 254:2716-2721.

Britten, R.J. 1972. DNA sequence interspersion and a speculation about evolution. *Brookhaven Symp. Biol.* 23:80-94.

Britten, R.J. & E.H.Davidson 1968. Gene regulation for higher cells: a theory. *Science* 165:349-357.

Britten, R.J. & E.H.Davidson 1971. Repetitive and nonrepetitive DNA sequences and a speculation on the origins of evolutionary noveltry. *Quart. Rev. Biol.* 46:111-133.

Britten, R.J. & E.H.Davidson 1976. DNA sequence arrangement and preliminary evidence on its evolution. *Fed. Proc.* 35:215-217.

Britten, R.J. & D.E.Kohne 1968. Repeated sequences in DNA. *Science* 161:529-540.

Britten, R.J., D.E.Graham, F.C.Eden, D.M.Painchaud & E.H.Davidson 1976. Evolutionary divergence and length of repetitive sequences in sea urchin DNA. *J. mol. Evol.* 9:1-23.

Britten, R.J., A.Cetta & E.H.Davidson 1978. The single-copy DNA sequence polymorphism of the sea urchin *Strongylocentrotus purpuratus. Cell* 15:1175-1186.

Broker, T.R. 1973. An electron microscopic analysis of pathways for bacteriophage T_4 DNA recombination. *J. mol. Biol.* 81:1-16.

Brykov, V.A., V.G.Volfson, E.G.Motrenko & V.I.Vorob'ev 1978. Single copy and repetitive DNA sequences of Echinodermata. I. Arrangement of DNA sequences of different multiplicity. *Mol. Biol. (Mosk.)* 12:1299-1312.

Brykov, V.A., V.G.Volfson & V.I.Vorob'ev 1979a. Single copy and repetitive nucleotide sequences in the genome of Echinodermata. II. Nucleotide sequence divergence of DNA of echinoderms. *Mol. Biol. (Mosk.)* 13:47-59.

Brykov, V.A., V.G.Volfson & V.I.Vorob'ev 1979b. Genome structure and divergence of nucleotide sequences in Echinodermata. *Chromosoma* 74:105-124.

Busslinger, M., R.Portmann & M.L.Birnsteil 1979. A regulatory sequence near the 3′ end of sea urchin histone genes. *Nuc. Acids Res.* 6:2997-3008.

Busslinger, M., R.Portmann, J.C.Irminger & M.L.Birnsteil 1980. Ubiquitous and gene-specific regulatory 5′ sequences in a sea urchin histone DNA clone coding for histone protein variants. *Nuc. Acids Res.* 8:957-977.

Chaudhari, N. 1980. Sequence organization and evolution of long and short repetitive DNA sequences in sea urchins. Ph.D. Dissert., University of South Carolina, Columbia.

Chaudhari, N. & S.P.Craig 1979a. The evolution of the long and short repetitive DNA sequences in sea urchins. *Biochem. Biophys. Acta* 562:438-452.

Chaudhari, N. & S.P.Craig 1979b. Internal organization of long repetitive DNA sequences in sea urchin genomes. *Proc. natl Acad. Sci. US* 76:6101-6105.

Cleveland, D.W., M.A.Lopata, R.J.MacDonald, R.J.Cowan, W.J.Rutter & M.W.Kirschner 1980. Number and evolutionary conservation of α- and β-tubulin and cytoplasmic α- and γ-actin genes using specific cloned cDNA probes. *Cell* 20:95-105.

Cohn, R.H. & L.H.Kedes 1979a. Nonallelic histone gene clusters of individual sea urchins *(Lytechinus pictus)*: polarity and gene organization. *Cell* 18:843-853.

Cohn, R.H. & L.H.Kedes 1979b. Nonallelic histone gene clusters of individual sea urchins *(L.pictus)*: mapping of homologies in coding and spacer DNA. *Cell.* 18:855-864.

Cohn, R.H., J.C.Lowry & L.H.Kedes 1976. Histone genes of the sea urchin *(S.purpuratus)* cloned in *E.coli:* order, polarity, and strandedness of the five histone-coding and spacer regions. *Cell* 9:147-161.

Craig, S.P., N.Chaudhari & M.Steinert 1979. Characterization of long and short repetitive sequences in the sea urchin genome. *Biochim. Biophys. Acta* 565:33-50.

Crain, W.R., E.H.Davidson & R.J.Britten 1976. Contrasting patterns of DNA sequence arrangement in *Apis mellifera* (Honeybee) and *Musca domestica* (Housefly). *Chromosoma* 59:1-12.

Davidson, E.H. & R.J.Britten 1979. Regulation of gene expression: repetitive sequences. *Science* 204:1052-1059.

Davidson, E.H., B.R.Hough, C.S.Amenson & R.J.Britten 1973a. General interspersion of repetitive with nonrepetitive sequence elements in the DNA of *Xenopus. J.molec. Biol.* 77:1-23.

Davidson, E.H., D.E.Graham, B.R.Neufeld, M.E.Chamberlin, C.S.Amenson, B.R.Hough & R.J.Britten 1973b. Arrangement and characterization of repetitive sequence elements in animal DNAs. *Cold. Spring Harbor Symp. quant. Biol.* 38:295-301.

Davidson, E.H., G.A.Galau, R.C.Angerer & R.J.Britten 1975a. Comparative aspects of DNA organization in Metazoa. *Chromosoma* 51:253-259.

Davidson, E.H., B.R.Hough, W.H.Klein & R.J.Britten 1975b. Structural genes adjacent to interspersed repetitive DNA sequences. *Cell* 4:217-238.

Davidson, E.H., T.L.Thomas, R.H.Scheller & R.J.Britten 1982. The sea urchin actin genes, and a speculation on the evolutionary significance of small gene families. In: G.Dover & R.B.Flavell (eds), *Genome Evolution.* Academic Press, London.

Doolittle, W.F. & C.Sapienza 1980. Selfish genes, the phenotype paradigm and genome evolution. *Nature, London* 284:601-603.

Durica, D.S., J.A.Schloss & W.R.Crain 1980. Organization of actin gene sequences in the sea urchin: molecular cloning of an introncontaining DNA sequence coding for a cytoplasmic actin. *Proc. natl Acad. Sci. US* 77:5683-5687.

Eden, F.C., D.E.Graham, E.H.Davidson & R.J.Britten 1977. Exploration of long and short repetitive sequence relationships in the sea urchin genome. *Nuc. Acids Res.* 4:1553-1567.

Edgell, M.H., F.Perler, N.Blin, C.A.Hutchinson & D.W.Stafford 1978. The ribosomal RNA gene region in the sea urchin *Lytechinus variegatus. Natl. Cancer Inst. Monogr.* 11-18.

Efstratiadis, A., J.W.Posakony, T.Maniatis, R.M.Lawn, C.O'Connell, R.A.Spritz, J.K.DeRiel, B.G.Forget, S.M.Weissmann, J.L.Slightom, A.E.Blechl, O.Smithies, F.E.Baralle, C.C.

Shoulders & N.J.Proudfoot 1980. The structure and evolution of the human β-globin gene family. *Cell* 21:653-668.
Eldredge, N. & S.J.Gould 1972. Punctuated equilibria: an alternative to phyletic gradualism. In: T.J.M.Schopf (ed.), *Models in Paleobiology*, pp.82-115. Freeman Cooper, San Francisco.
Farguhar, M.N. & B.J.McCarthy 1973. Evolutionary stability of the histone genes of sea urchins. *Biochemistry* 12:4113-4122.
Flavell, R., J.Rimpau, D.B.Smith, M.O'Dell & J.R.Bedbrook 1980. The evolution of plant genome structure. In: C.L.Leaver (ed.), *Genome Organization and Expression in Plants*, pp.35-47. Plenum Press, New York.
Galau, G.A., M.E.Chamberlin, B.R.Hough, R.J.Britten & E.H.Davidson 1976a. Evolution of repetitive and nonrepetitive DNA. In: F.J.Ayala (ed.), *Molecular Evolution*, pp. 200-224. Sunderland, Mass., Sinauer Associates.
Galau, G.A., W.H.Klein, M.M.Davis, B.J.Wold, R.J.Britten & E.H.Davidson 1976b. Structural gene sets active in embryos and adult tissues of the sea urchin. *Cell* 7:487-505.
Georgiev, G.P. 1969. On the structural organization of operon and the regulation of RNA synthesis in animal cells. *J. theor. Biol.* 25:473-490.
Goldberg, R.B., G.A.Galau, R.J.Britten & E.H.Davidson 1973. Nonrepetitive DNA sequence representation in sea urchin embryo messenger RNA. *Proc. natl Acad. Sci. US* 70: 3516-3520.
Gross, K., W.Schaffner, J.Telford & M.Birnstiel 1976. Molecular analysis of the histone gene cluster of *Psammechimus miliaris:* III. Polarity and asymmetry of the histone-coding sequences. *Cell* 8:479-484.
Grula, J.W., T.J.Hall, J.A.Hunt, T.D.Giugni, G.J.Graham, E.H Davidson & R.J.Britten 1982. Sea urchin DNA sequence variation and reduced interspecies differences of the less variable DNA sequences. *Evolution* 36:665-676.
Hall, T.J., J.W.Grula, E.H.Davidson & R.J.Britten 1980. Evolution of sea urchin nonrepetitive DNA. *J. molec. Evol.* 16:95-110.
Harpold, M.M. & S.P.Craig 1977. The evolution of repetitive DNA sequences in sea urchins. *Nuc. Acids Res.* 4:4425-4437.
Harpold, M.M. & S.P. Craig 1978. The evolution of nonrepetitive DNA in sea urchins. *Differentiation* 10:7-11.
Hentschel, C.C. & M.L.Birnstiel 1981. The organization and expression of histone gene families. *Cell* 25:301-313.
Hinegardner, R. 1974. Cellular DNA content of the Echinodermata. *Comp. Biochem. Physiol.* 49:219-226.
Jagadeesivaran, P., B.G.Forget & S.M.Weissman 1981. Short interspersed repetitive DNA elements in eucaryotes: Transposable DNA elements generated by reverse transcription of RNA Pol. III transcripts? *Cell* 26:141-142.
Joseph, D.R. & D.W.Stafford 1976. Purification of sea urchin ribosomal RNA genes with a single-strand specific nuclease. *Biochim. Biophys. Acta* 418:167-174.
Kedes, L.H. 1979. Histone genes and histone messengers. *Ann. Rev. Biochem.* 48:837-870.
Kedes, L.H. & M.L.Birnstiel 1971. Reiteration and clustering of DNA sequences complementary to histone messenger RNA. *Nature, London* 230:165-169.
Kedes, L.H., A.C.Chang, D.Houseman & S.N.Cohen 1975. Isolation of histone genes from unfractionated sea urchin DNA by subculture cloning in *E.coli. Nature, London* 255: 533-538.
Klein, W.H., T.L.Thomas, C.Lai, R.H.Scheller, R.J.Britten & E.H.Davidson 1978. Characteristics of individual repetitive sequence families in the sea urchin genome studied with cloned repeats. *Cell* 14:889-900.

Leder, P., J.N.Hansen, D.Konkel, A.Leder, Y.Nishioka & C.Talkington 1980. Mouse globin system: a functional and evolutionary analysis. *Science* 209:1336-1342.

Lee, A.S., R.J.Britten & E.H.Davidson 1977. Short-period repetitive sequence interspersion in cloned fragments of sea urchin DNA. *Cold Spring Harbor Symp. quant. Biol.* 42: 1065-1076.

Lewin, B. 1980. *Gene Expression, Vol. 2, Eucaryotic Chromosomes* (2nd ed.). 1160pp. John Wiley & Sons, New York.

Lu, A.L., D.A.Steege & D.W.Stafford 1980. Nucleotide sequence of a 5s ribosomal RNA gene in the sea urchin *L.variegatus. Nuc. Acids. Res.* 8:1839-1853.

Manning, J.E., C.W.Schmid & N.Davidson 1975. Interspersion of repetitive and nonrepetitive DNA sequences in the *Drosophila melanogaster* genome. *Cell.* 4:141-155.

Mellado, R.P., H.Delius, B.Klein & K.Murray 1981. Transcription of sea urchin histone genes in *Escherichia coli. Nuc. Acids Res.* 9:3889-3906.

Miklos, G.L.G. & A.C.Gill 1981. The DNA sequences of cloned complex satellite DNAs from Hawaiian *Drosophila* and their bearing on satellite DNA sequence conservation. *Chromosoma* 82:409-427.

Mishra, N.R. 1979. Study of ribosomal deoxyribonucleic acid of sea urchin. *Mol. Biol. Rep.* 4:241-246.

Moore, G.P., R.H.Scheller, E.H.Davidson & R.J.Britten 1978. Evolutionary change in the repetition frequency of sea urchin DNA sequences. *Cell* 15:649-660.

Moore, G.P., F.D.Constantini, J.W.Posakony, E.H.Davidson & R.J.Britten 1980. Evolutionary conservation of repetitive sequence expression in sea urchin egg RNA's. *Science* 208:1046-1048.

Moore, G.P., W.R.Pearson, E.H.Davidson & R.J.Britten 1981. Long and short repeats of sea urchin DNA and their evolution. *Chromosoma* 84:19-32.

Moyzis, R.K., J.Bonnet, D.W.Li & P.O.P.Ts'o 1981. An alternative view of mammalian DNA sequence organization II. Short repetitive sequences are organized into scrambled tandem clusters in Syrian Hamster DNA. *J. mol. Biol.* 153:871-896.

Musti, A.M., D.A.Sobieski, B.B.Chen & F.C.Eden 1981. Repeated deoxyribonucleic acid clusters in the chicken genome contain homologous sequence elements in scrambled order. *Biochem.* 20:2989-2999.

Nijhawan, P. & W.F.Marzluff 1979. Metabolism of low molecular weight ribonucleic acids in early sea urchin embryos. *Biochem.* 18:1353-1360.

Ohta, T. 1981. Population genetics of selfish DNA. *Nature, London* 292:648-649.

Orgel, L.E. & F.H.C.Crick 1980. Selfish DNA: the ultimate parasite. *Nature, London.* 284:604-607.

Overton, G.C. & E.S.Weinberg 1978. Length and sequence heterogeneity of the histone gene repeat unit of the sea urchin, *S.purpuratus. Cell* 14:247-257.

Patterson, J.B. & D.W.Stafford 1970. Sea urchin satellite deoxyribonucleic acid. Its large-scale isolation and hybridization with homologous ribosomal ribonucleic acid. *Biochemistry* 9:1278-1283.

Pearson, W.R. & J.F.Morrow 1981. Discrete-length repeated sequences in eukaryotic genomes. *Proc. natl Acad. Sci. US* 78:4016-4020.

Perler, F., A.Efstratiadis, P.Lomedico, W.Gilbert, R.Kolodner & J.Dodgson 1980. The evolution of genes: the chicken preproinsulin gene. *Cell* 20:555-565.

Petrov, N.B., A.B.Poltaraus & A.S.Antonov 1980. Determination of the degree of ribosomal genes conservativity in some invertebrate species. *Biokhimiia* 45:165-172.

Poltaraus, A.B., A.S.Antonov & N.B.Petrov 1980a. Divergence of repetitive DNA sequences in Echinodermata. I. Comparison of sequences with a high degree of intragenomic divergence. *Mol. Biol. (Mosk.)* 14:661-674.

Poltaraus, A.B., N.B.Petrov & A.S.Antonov 1980b. Divergence of repetitive DNA sequences in echinoderms. II. Comparison of sequences with a low degree of intragenome divergence. *Mol. Biol. (Mosk.)* 14:1046-1056.

Portmann, R., W.Schaffner & M.Birnstiel 1976. Partial denaturation mapping of cloned histone DNA from the sea urchin *Psammechinus miliaris. Nature, London* 264:31-34.

Robertson, H.D. & E.Dickson 1975. RNA processing and the control of gene expression. *Brookhaven Symp. Biol.* 26:240-266.

Royal, A., A.Garapin, B.Cami, F.Perin, J.L.Mandel, M.LeMeur, F.Bregegere, F.Gannon, J.P.LePennec, P.Chambon & P.Kourilsky 1979. The ovalbumin gene region: common features in the organization of three genes expressed in chicken oviduct under hormonal control. *Nature, London* 279:125-132.

Schaffner, W., K.Gross, J.Telford & M.Birnstiel 1976. Molecular analysis of the histone gene cluster of *Psammechinus miliaris.* II. The arrangement of the five histone-coding and spacer sequences. *Cell* 8:1471-1478.

Schaffner, W., G.Kunz, H.Daetwyler, J.Telford, H.O.Smith & M.L.Birnstiel 1978. Genes and spacers of cloned sea urchin histone DNA analyzed by sequencing. *Cell* 14:655-671.

Scheller, R.H., L.B.McAllister, W.R.Crain jr, D.S.Durica, J.W.Posakony, T.L.Thomas, R.J. Britten & E.H.Davidson 1981a. Organization and expression of multiple actin genes in the sea urchin. *Molec. cell. Biol.* 1:609-628.

Scheller, R.H., D.M.Anderson, J.W.Posakony, L.B.McAllister, R.J.Britten & E.H.Davidson 1981b. Repetitive sequences of the sea urchin genome II. Subfamily structure and evolutionary conservation. *J. mol. Biol.* 149:15-39.

Schuler, M.A & E.B.Keller 1981. The chromosomal arrangement of two linked actin genes in the sea urchin *S.purpuratus. Nuc. Acids Res.* 9:591-604.

Smith, M.J. & R.Boal 1978. DNA sequence organization in the common Pacific starfish *Pisaster ochraceus. Can. J. Biochem.* 56:1048-1054.

Smith, M.J., A.Luis, K.K.Gibson & J.K.Etzkorn 1980. DNA sequence organization in the starfish *Dermasterias imbricata. Can. J. Biochem.* 58:352-360.

Smith, M.J., R.Nicholson, M.Stueryl & A.Lui 1982. Single copy DNA homology in sea stars. *J. mol. Evol.* 18:92-101.

Sobieski, D.A. & F.C.Eden 1981. Clustering and methylation of repeated DNA: persistence in avaian development and evolution. *Nuc. Acids Res.* 9:6001-6016.

Sy, J. & K.S.McCarty 1970. Characterization of 5.8-S RNA from a complex with 26-S ribosomal RNA from *Arbacia punctulata. Biochim. Biophys. Acta* 199:86-94.

Vorob'ev, V.I. & G.N.Kosjuk 1974. Distribution of repetitive and non-repetitive nucleotide sequences in the DNA of sea urchin. *F.E.B.S. Lett.* 47:43-46.

Wahli, W., I.B.Dawid, T.Wyler, R.Weber & G.U.Ryffel 1980. Comparative analysis of the structural organization of two closely related vitellogenin genes in *X.leavis. Cell* 20: 107-117.

Weinberg, E.S., M.L.Birnstiel, I.F.Purdom & R.Williamson 1972. Genes coding for polysomal 9s RNA of sea urchins: conservation and divergence. *Nature, London* 240:225-228.

Weinberg, E.S., G.C.Overton, R.H.Shutt & R.H.Reedes 1975. Histone gene arrangement in the sea urchin, *Strongylocentrotus purpuratus. Proc. natl Acad. Sci. US* 72:4815-4819.

Weinblum, D., U.Gungerich, M.Geisert & R.K.Zahn 1973. Occurrence of repetitive sequences in the DNA of some marine invertebrates. *Biochim. Biophys. Acta* 299:231-240.

Wensink, P.C., S.Tabata & C.Pachl 1979. The clustered and scrambled arrangement of moderately repetitive elements in *Drosophila* DNA. *Cell* 18:1231-1246.

Wilson, R.E., N.Blin & D.W.Stafford 1976. A denaturation map of sea urchin ribosomal DNA. *Chromosoma* 58:247-253.

Zuckerkandl, E. 1980. Sequences, phenotypes and directional evolution. In: *Second International Congress of Systematic and Evolutionary Biology*, p.124. Univ. of British Colombia, Vancouver.

RESPIRATORY GAS EXCHANGE IN ECHINODERMS

J.MALCOLM SHICK
Department of Zoology, University of Maine, Orono, USA

CONTENTS

1 INTRODUCTION

Echinoderms have received little attention from respiratory physiologists compared to other invertebrate phyla. This stems in part from their size and shape, which often are not conducive to placement in respirometers, and it is thus no coincidence that the relatively compact echinoids are best represented in Lawrence & Lane's (1982) compilation of oxygen uptake rates in the phylum. Moreover, most echinoderms do not exhibit measurable amounts of convection of the external medium or perfusion of the respiratory surface, and lack respiratory pigments, all features of some interest to physiologists. Holothuroids provide an exception to these generalizations and are therefore reasonably well represented in the original physiological literature and texts. Still, the prevailing view seems to be that echinoderms in general are at the mercy of their respiratory environment, possessing few adaptations for external and internal transport of oxygen relative to other higher invertebrates. Their rates of oxygen uptake are generally lower than in other invertebrates, probably because of their lesser degree of muscular development (Lawrence & Lane 1982). Even their high 'oxyregulatory' capacity is seen as a passive consequence of the isolation of much of their biomass within a massive protective envelope (Mangum & Van Winkle 1973).

Within the Echinodermata, comparison of oxygen uptake rates is difficult because they are usually expressed in terms of wet or dry weight, and the levels of inorganic components may vary widely, both inter- and intraspecifically. Even rates expressed in terms of total organic matter, organic nitrogen, or protein are problematic owing to differences in skeletal plans, body wall morphology, and degree of development of muscles and internal tissues. Perusing the oxygen uptake rates that have been reported in echinoderms (Lawrence & Lane 1982, Table 1), one is tempted to conclude, as did Farmanfarmaian 15 years ago, that 'under approximately normal conditions the respiratory rate is of the same order of magnitude for the phylum as a whole . . .'

Nevertheless, some insight into respiratory adaptations among the classes of Echinodermata can be gained at the organismal level. The equation $\dot{V}_{O_2} = aW^b$ describes the relationship between the rate of oxygen uptake ($\dot{V}_{O_2}$) and body mass (W), where a is the intercept and b is the slope of the regression of log $\dot{V}_{O_2}$ on log W.* As $\dot{V}_{O_2}$ is not directly proportional to body mass (that is, $b < 1.0$), the mass-specific $\dot{V}_{O_2}$ is greater in smaller individuals. Values of b have been empirically determined in a number of echinoderms (summarized in Table 2 in Lawrence & Lane 1982). Informatively, b is greatest (mean value of 0.85) in the holothuroids, in which the respiratory trees deliver oxygen directly to the interior milieu. Conversely, the smaller b (0.64, which approaches the expected value of 0.67 if $\dot{V}_{O_2}$ were proportional to body surface area) in the echinoids seems related to their often nearly spherical shape and calcified test, both of which tend to insulate internal tissues from external oxygen. The high value (0.84) in asteroids, and low value (0.54) in ophiuroids, are correlated with the relatively flattened body form, large total surface area of tube feet, and presence of numerous thin-walled papulae in the former, and an increasing proportion of metabolically inert inorganic material with increasing body size in the latter.

* Symbols and abbreviations:
P_{O_2}: partial pressure of oxygen; P_{CO_2}: partial pressure of carbon dioxide; Pi_{O_2}: P_{O_2} of incurrent ventilatory fluid; Pe_{O_2}: P_{O_2} of excurrent ventilatory fluid; Ppf_{O_2}: P_{O_2} of perivisceral coelomic fluid; ΔP_{O_2}: $Pe_{O_2} - Ppf_{O_2}$; C_{O_2}: concentration of oxygen in a fluid; C_{CO_2}: concentration of total carbon dioxide in a fluid;
$\dot{V}_{O_2}$: rate of oxygen uptake; $\dot{V}w$: rate of water flow across the site of oxygen uptake; $\frac{\dot{V}w}{\dot{V}_{O_2}}$: convection requirement; $\%Extr_w$: $\frac{Pi_{O_2} - Pe_{O_2}}{Pi_{O_2}} \times 100$, percent extraction or utilization of oxygen from the ventilatory fluid; $\%Ew_{O_2}$: $\frac{Pi_{O_2} - Pe_{O_2}}{Pi_{O_2} - Ppf_{O_2}} \times 100$, effectiveness of extraction of oxygen from the ventilatory fluid, i.e. actual oxygen uptake as percent of maximum possible, $\dot{Q}_H$: rate of heat production, $\dot{Q}_{OX}$: caloric equivalent of oxygen uptake, calculated as $20.3\ J \cdot ml\ O_2^{-1}$.

This superficial analysis suggests that a comparative approach, long favored by anatomists and paleontologists, and increasingly by physiologists, will be enlightening when applied to the Echinodermata.

Earlier reviewers of this topic (Hyman 1955, Farmanfarmaian 1966, Binyon 1972) have enumerated the principal respiratory surfaces and mechanisms in echinoderms, and these will not be repeated here. Rather, particular structural adaptations will be discussed in terms of their physiological manifestations. At the respiratory surface, the parameters of interest are contained in the modified Fick equation for diffusive transfer (Dejours 1979): $\dot{V}_{O_2} = K_{O_2} \cdot \frac{A}{E} \cdot \Delta P_{O_2}$, where $\dot{V}_{O_2}$ is the rate of oxygen uptake, K_{O_2} is Krogh's diffusion constant (the quantity of oxygen diffusing per unit time through a unit cross-sectional area under a unit partial pressure gradient) and is a property of the substance through which oxygen diffuses, A is the surface area across which diffusion occurs, E is the length of the diffusion path, and ΔP_{O_2} is the difference in partial pressure of oxygen between the two locations. (A similar equation applies to CO_2 transfer.) Since the maximum value of ΔP_{O_2} is normally limited by the atmospheric P_{O_2} of about 155 mm Hg, anatomical adaptations at the respiratory surface involve decreases in E, increases in A, and perhaps increases in K_{O_2} (by changing the proportions of muscular and connective tissue, which have slightly different water contents and diffusive resistances). Ironically, mucus, which has a high diffusive resistance (low K_{O_2}), nevertheless has an important respiratory role in cleansing external gas exchange surfaces and stabilizing burrows in infaunal forms.

The fundamental relationship $\frac{\dot{V}w}{\dot{V}_{O_2}} \cdot Ci_{O_2} \cdot \%Extr_w = 1.0$ (Dejours 1975), where $\dot{V}w$ is the water flow rate, Ci_{O_2} is the oxygen concentration of the incurrent water, and $\%Extr_w$ is the percent extraction of oxygen from the water, enables the calculation of any one variable when the other three are known. Although this is of little relevance in epifaunal echinoderms that do not create localized respiratory currents, it is quite useful in holothuroids having tidally-ventilated respiratory trees and in forms which irrigate burrows, in which cases $\dot{V}w$ or $\%Extr_w$ can be measured directly. The corresponding equation for blood flow likewise has little relevance in most echinoderms, in which most of the transport of respiratory gases is not confined to a closed vascular circulation. However, it has been applied to oxygen transported by the fluid in the tube foot-ampullar system (Colacino 1973).

The present paper concerns whole-organism aspects of respiratory physiology in echinoderms, and concentrates on advances made since the earlier reviews. As cellular respiration has been reviewed by Ellington (1982) and rates of oxygen uptake and exogenous factors affecting them have been reviewed by Lawrence & Lane (1982), these aspects will not be considered here, except as they are related to organismal or systemic responses.

Their large body size itself dictates that echinoderms must have special surfaces and mechanisms to promote oxygen delivery to their internal tissues, and earlier reviewers of this topic (Hyman 1955, Farmanfarmaian 1966, Binyon 1972) catalogued these adaptations. In the interim increasing attention has been paid to the functional anatomy and morphology of gas exchange systems and these advances will be treated here. This topic has been investigated and reviewed by Paul (1968, 1972, 1976, 1977) for fossil echinoderms.

Classically, the involvement of a structure in oxygen uptake by an intact echinoderm was investigated by covering the structure (tube feet of asteroid and echinoids), ablating it (peristomial gills of echinoids), or occluding the opening to it (cloaca of holothuroids), and observing the effects on $\dot{V}_{O_2}$. Pentreath (1971) confirmed the long-suspected importance of the bursae in ophiuroids as respiratory organs: occlusion of the bursae in *Ophionereis fasciata* and *Ophiopteris antipodum* resulted in an average 27 % decline in $\dot{V}_{O_2}$. A gas exchange role was also inferred from the ramifications of the very thin (thinner than tube foot walls) bursal membrane within the disk, providing a large surface with little diffusive resistance for oxygen delivery to the coelom.

Cole & Burggren (1981) found that oxygen uptake via either the aboral surface (especially across the papulae) or the oral surface exceeded 50 % of the total $\dot{V}_{O_2}$ in *Asterias forbesi.* They suggested that oxygen uptake across one surface is increased when the other surface is deprived of oxygen, perhaps due to an increased diffusion gradient caused by a decline in perivisceral fluid P_{O_2} when one respiratory surface is covered. The greater $\dot{V}_{O_2}$ associated with the oral surface may also be due to a large oxygen consumption by the tube feet.

There is little else to add to earlier descriptions of respiratory surfaces in asteroids. The apical or 'epiproctal' cone in burrowing forms such as *Astropecten, Ctenodiscus,* and the porcellanasterids certainly has a respiratory role. The extension of the cone greatly reduces its wall thickness and thus the diffusion distance between seawater and the coelomic fluid; moreover, the peristalsis by the cone musculature enhances mixing of the internal fluid, particularly under hypoxic conditions (Shick 1976). The principal respiratory organs are the papulae, even in the Porcellanasteridae (Madsen 1961), which erroneously had been reported to lack them (Hyman 1955). The clustering of papulae into papularia on the proximal midradial surface of the arms in phanerozoids such as *Benthopecten* and *Pectinaster* may concentrate circulation of oxygenated perivisceral fluid directly across the gonads and pyloric caeca in these flattened forms.

Colacino (1973) attempted to model the respiratory contribution of the tube foot-ampullar gas exchange system, which accounts for about 40 % of total $\dot{V}_{O_2}$, in the holothuroid *Sclerodactyla briareus.* In so doing he provided important primary data on the morphometry of the system and on the dynamics of flow within it. As in other echinoderms, the wall of the ampulla is thinner than that of the tube foot, and may facilitate transfer of oxygen from the ampulla to the perivisceral fluid. Flow is simultaneously bidirectional within a single tube foot and ampulla, despite the lack of a partition or septum, which implies that the direction of ciliary beating is opposite on the opposing inner walls of the system. Thus, a dual-pore entrance and exit (as in the echinoids) is not a prerequisite for continuous-flow perfusion of the tube feet. However, the lack of a partition between currents flowing into and out of a tube foot allows their mixing or at least diffusion of gases between the streams. As a result, the two streams tend to equilibrate over short distances, which reduces the difference between them in P_{O_2} and thus limits the amount of oxygen delivered to the interior.

Mean velocity of water vascular fluid flow in either stream within a tube foot of *S.briareus* is 0.0195 $cm \cdot s^{-1}$, and mean volume ($\dot{V}$) of flow is 4.446 x $10^{-6} ml \cdot s^{-1}$; mean maximum values for velocity and $\dot{V}$ are 0.039 $cm \cdot s^{-1}$ and 8.893 x $10^{-6} ml \cdot s^{-1}$, respectively. The model indicates that changes in flow velocity have a greater effect on the amount of oxygen delivered than do changes in tube foot wall thickness. Despite the uncertain function of the holothuroid's water vascular hemoglobin *in vivo*, the predictions of the model are in agreement with the measured rates of oxygen uptake across the general body surface (see section 5).

Engstrom (1982) suggested that the water vascular hemoglobin in the dendrochirote *Cucumaria lubrica* may transport oxygen to eggs and embryos brooded between the female and the substratum, where water circulation is minimal or absent and where P_{O_2} is low. Whereas no adult holothuroid had a red coloration prior to the breeding season, more than half of the brooding females had a red appearance on their ventral surfaces.

Investigation of the other component of the dual gas exchange system in holothuroids, the respiratory trees, has been more functionally than structurally oriented. Considering the differences in habitat (infaunal *vs.* epifaunal), morphology (varying extents of body surface covered by tube feet), and anatomy (varying dimensions of tube feet) among aspidochirotes, dendrochirotes, and molpadiids, the degree of development of the respiratory trees might also vary systematically among them, and a comparative study of the morphometry of the holothurian water lung could prove interesting. For example, respiratory tree weight scales as the 0.967 power of body weight in *Cucumaria frondosa* and the 1.131 power in *Sclerodactyla briareus* (Brown 1978). These exponents are significantly different, and the greater increase in respiratory tree size with body weight in *S.briareus* may be related to this species' occupation of a more stressful (muddy infaunal) respiratory environment.

Conversely, the complex blood vascular system with its multiple hearts, unidirectional flow, and extensive respiratory plexus associated with the left respiratory tree in aspidochirotes, especially *Stichopus moebii* (Herreid *et al.*

Table 1. P_{O_2} and P_{CO_2} in echinoderm perivisceral fluids taken from specimens near air-saturation. All data are for laboratory-maintained animals, except as indicated. Error terms are ± 1 S.E., except in data of J.A.Petersen, where ± 1 S.D. is used. Reproductive condition (spent or ripe) is indicated where known.

Species	Ambient P_{O_2} (mm Hg)	Ppf_{O_2} (mm Hg)	Ppf_{CO_2} (mm Hg)	Temp. (°C)	Reference
Asteroidea					
Asterias rubens	150	84 ± 4		15	Shick, unpublished
Asterias vulgaris	155	101 ± 4^{sp}		21	Shick, unpublished
Astropecten brasiliensis	150	86 ± 4	1.0 ± 0.1	23-25	Petersen, pers. comm.
	130-140	40 ± 6^{s}	1.5 ± 0.1	23-25	
Ctenodiscus crispatus	159	135		5	Shick *et al.* 1981
Echinaster brasiliensis	130	53 ± 2	1.1 ± 0.1	23-25	Petersen, pers. comm.
Echinaster sepositus	151	85 ± 3		25	Houlihan & Duthie 1981
	151	$66^{f,n}$-$75^{f,d}$		25	
Enoplopatiria stellifera	110-135	46 ± 11	1.5 ± 0.3	23	Petersen, pers. comm.
Luidia clathrata	135-145	65 ± 2	2.2 ± 0.1	23-25	Petersen, pers. comm.
Marthasterias glacialis	150	104 ± 5^{sp}		15	Shick, unpublished
Pteraster tesselatus	145		1.00-1.40	11-12	Johansen & Petersen 1971
	145	105		10	
Echinoidea					
Allocentrotus fragilis	150	115-120		8	Webster 1975
Arbacia lixula	151	87 ± 3		25	Houlihan & Duthie 1981
	151	$87^{f,d}$-$96^{f,n}$		25	
Arbacia lixula	125-140	55 ± 15	1.6 ± 0.6	21-23	Petersen, pers. comm.
	130	50 ± 11	1.4 ± 0.2	26	
	145-160	53 ± 13^{f}	1.7 ± 0.3	25-27	
Clypeaster subdepressus	120	86 ± 6^{s}	2.2 ± 0.2	23-25	Petersen, pers. comm.
Echinocardium pennatifidum	147-155	55 ± 8^{r}		15	Shick, unpublished
Echinometra lucunter	125-140 (stirred)	73 ± 11	1.6 ± 0.2	21-23	Petersen, pers. comm.
	125-140 (unstirred)	55 ± 7	2.6 ± 0.2	21-23	
	140-155	66 ± 11^{f}	2.8 ± 0.4	26	
Echinus esculentus	147-155	80 ± 3^{sp}		15	Shick, unpublished
	147-155	37 ± 5^{r}		15	
	147-155	$91 \pm 4^{pe,sp}$		15	

	147-155	$48 \pm 2^{pe,r}$		15	
Encope emarginata	100-120	67 ± 12	3.3 ± 0.7	23-25	Petersen, pers. comm.
	135-140	96 ± 15^{f}	2.6 ± 0.9	23-25	
Lytechinus variegatus	125-145	69 ± 13	1.8 ± 0.5	23-25	Petersen, pers. comm.
	120-155	58 ± 11^{f}	2.1 ± 0.3	23-25	
Paracentrotus lividus	151	94 ± 4		25	Houlihan & Duthie 1981
	151	$56^{f,n}$-$89^{f,d}$		25	
Spatangus purpureus	155	86^{sp}		15	Shick, unpublished
	147-155	57 ± 2^{r}		15	
Sphaerechinus granularis	151	84 ± 3		25	Houlihan & Duthie 1981
	151	81 ± 4^{f}		25	
Strongylocentrotus droebachiensis	155	13		6	Steen 1965
	155	26		16	
Strongylocentrotus droebachiensis	135	58		9.5-10.5	Johansen & Vadas 1967
Strongylocentrotus droebachiensis	145	100		13	Sabourin & Stickle 1981
Strongylocentrotus droebachiensis	155	100 ± 2^{sp}		10	J.A.Dykens, pers. comm.
Strongylocentrotus franciscanus	135	97-104		10	Johansen & Vadas 1967
Strongylocentrotus purpuratus	135	60-85		9.5-10.5	Johansen & Vadas 1967
	130	2-90^{f}		9.5-10.5	
Strongylocentrotus purpuratus	150	45		13	Webster 1975
Strongylocentrotus purpuratus	150	23-80		13 ± 3	Webster & Giese 1975
Holothuroidea					
Cucumaria frondosa	140	102		10	Brown & Shick 1979
Eupentacta quinquesemita	145	88 ± 7		13	Sabourin & Stickle 1981
Holothuria forskali	131	131		17	Newell & Courtney 1965
Holothuria forskali	150	88 ± 6		15	Shick, unpublished
Holothuria grisea	130-140	73 ± 8	1.1 ± 0.3	23	Petersen, pers. comm.
	145-150	54 ± 6^{f}	1.1 ± 0.2	23	
Holothuria polii	151	79 ± 5		25	Houlihan & Duthie 1981
	151	81 ± 4^{f}		25	
Isostichopus badionotus	135-140	71 ± 9		23	Petersen, pers. comm.
Molpadia arenicola	140-150	80	1.2		K.Johansen, pers. comm.
Sclerodactyla briareus	159	122		21-23	Mangum & Van Winkle 1973

d: Sample taken during day; f: Sample taken in field; n: Sample taken at night; pe: Periesophageal coelomic fluid sample; r: Specimens with ripe gonads; sp: Reproductively spent specimens; s: Specimens buried in sediment.

1976), is anatomically well-known, but its quantitative role in gas exchange remains unevaluated. The lack of blood vascular hemoglobin in this species may limit the gas exchange capacity of the system but would simplify its analysis.

The lack of respiratory trees in some holothuroids has necessitated other adaptations to enhance oxygen delivery. These include a thinning of the body wall in synaptids (Mangum & Van Winkle 1973), a gelatinous consistency brought about by a high water content in deep-sea elasipodids, and increases in surface:volume ratios in both of these groups, achieved through extreme vermiformicity in synaptids and by the bizarre dorsal papillae in some elasipodids (Hansen 1975). Although the intradermal ambulacral cavities in elasipodids may be an adaptation for walking, the pumping of the cavities may also increase respiratory exchange (Hansen 1975).

Knowledge of structural features to promote gas exchange has advanced most in the echinoids. *A priori,* it is in these sub-spherical, heavily plated forms that one would expect diffusion to be most limiting, particularly as body mass increases, and where adaptations to overcome this 'suffocation effect' (Paul 1977) would be anticipated.

The presence of peristomial gills in most regular echinoids has long been known, although the failure of their ablation to affect whole-organismal $\dot{V}_{O_2}$ in *Strongylocentrotus purpuratus* suggests no real respiratory role (Farmanfarmaian 1966). The prevailing view is that these distensible evaginations of an otherwise rigid body wall compensate for pressure changes in the peripharyngeal coelom caused by feeding activity of Aristotle's lantern (Kier 1974), but that they may have the secondary function of providing oxygen to the lantern muscles, which based on their very low lactate dehydrogenase, low phosphoenolpyruvate carboxykinase, and high pyruvate kinase activities (Zammit & Newsholme 1976, Bianconcini *et al.* 1980) appear to be more aerobic than most internal tissues (see section 6). Data on *Echinus esculentus* support this suggestion: the P_{O_2} of the peripharyngeal coelomic fluid is about 11 mm Hg higher than that in the main perivisceral coelomic fluid (Table 1).

The echinoid tube foot has received considerable attention from anatomists, and functional adaptations to promote gas exchange in the tube foot-ampullar system have been described throughout the class. Paired ambulacral pores provide a physical basis for separate incurrent and excurrent flows to the tube feet and ampullae, unlike the situation in *Sclerodactyla briareus.* The presumed respiratory role of various pore structures in extinct cystoids has been studied by Paul (1968, 1972, 1976).

Smith (1978) has separated the tube feet of regular echinoids into two broad categories according to the structure of their pores: P-type (partitioned isopores) and C-type (conjugate isopores). Within these groups can be found tube feet highly specialized for gas exchange (P1 and C1 tube feet). In P1 tube feet (Fig.1a), the pores through the test are rather large, the walls are thin (1 to

10 μm), and there is a well-developed septum extending roughly three-fourths of the length of the lumen of the tube foot. These factors, respectively, allow a large, one-way flow through the tube foot, minimize diffusive resistance between the lumen and the ambient seawater, and channel and separate the incurrent and excurrent streams. The columnar shape of the tube feet, the necessarily bidirectional, vertical flow within them, and the movement of water across the test mean that the internal and external flows are cross-current to each other (Fig.1a), allowing for reasonably efficient gas exchange (Dejours 1975, p.84). the efficiency may be increased by the collagenous septum, which reduces stream-stream diffusion, particularly near the entry and exit points of the two currents. In P2 tube feet (Fig.1b), the separation of incurrent and excurrent streams is also effected by the inner muscle layer, which is constricted near the septum, creating a 'dumbbell' shape in cross-section (Fig.1c) and thereby reducing the area of apposition of the two streams.

C1 tube feet, which occur in arbacioids and cidaroids, are more complex and consist of two thin-walled (ca.2 μm) cylinders which communicate via a flattened central convoluted area which has many septa (Fig.1e). The convoluted area greatly increases the surface area of the tube foot, divides the deoxygenated incurrent fluid into many streams (which flow counter to the external current), and provides a more complete separation of incurrent and excurrent flows than in P-type tube feet. Presumably, therefore, C1 tube feet are more efficient in gas exchange than are P1 tube feet. Both C1 and P1 tube feet typically occur aborally, a position which places them away from the substrate and into the oxygenated water column (Fenner 1973, Smith 1978).

Although *Strongylocentrotus droebachiensis* lacks highly specialized C1 and P1 tube feet, oxygen delivery is increased by a multiserial arrangement of its less efficient P2 tube feet (Smith 1978), an arrangement which also increases the strength of attachment by the echinoid. Steen (1976) attempted to quantify the gas exchange capacity in this species using a modified Fick diffusion equation and a diffusion constant for connective tissue. He measured the surface area provided by a 70-g individual's 1000 tube feet, their wall thickness, and the P_{O_2} of the perivisceral fluid and the external medium; substituting these values in the equation gave a theoretical respiratory capacity of 19 ml $O_2 \cdot h^{-1}$, well in excess of the measured $\dot{V}_{O_2}$ of 2 ml $O_2 \cdot h^{-1}$ in a 70-g specimen. Steen recognized that he may have: overestimated the area provided by the tube feet by assuming an equal amount of extension for all of them; slightly underestimated their wall thickness; and overestimated the diffusion gradient (which seems likely, in view of additional data in Table 1). Still, the theoretical respiratory capacity would remain much greater than the actual value. As Colacino (1973) showed, the most serious error was assuming that oxygen uptake occurred along the entire 20 mm length of a tube foot. Using a value of 4 mm for equilibration path length predicted by Colacino's model reduces the calculated respiratory capacity to less than 4 ml $O_2 \cdot h^{-1}$. It may not be purely coincidental that the length of

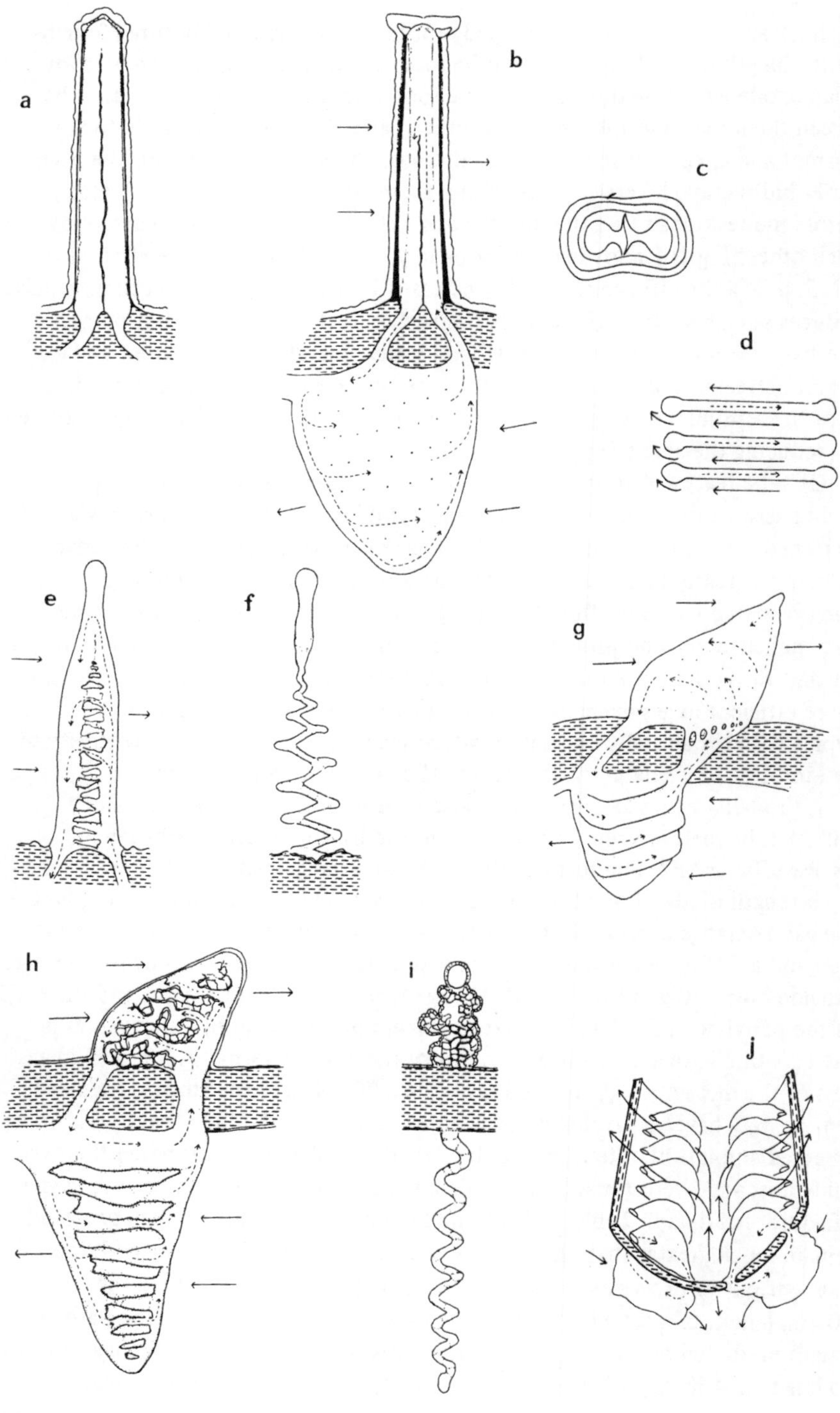
a
b
c
d
e
f
g
h
i
j

the septum separating the incurrent and excurrent streams in a tube foot of *Strongylocentrotus* is about 4 mm (Fig.2 in Fenner 1973).

Ampullae associated with tube feet showing respiratory adaptations typically are long, flattened and thin-walled. The flattened shape further separates the oxygenated current entering the ampulla from the deoxygenated current leaving it; within the ampulla the ciliary currents are channelled in parallel by septa, presumably promoting laminar flow along the exchange surface. Flattening also allows a flow of fluid within the ampulla over a long distance at right angles to the incurrent and excurrent streams thereby enabling the establishment of a ciliary counter-current in the perivisceral fluid along the exterior of the ampulla. Such a counter-current arrangement of ampullar internal and external flows occurs in the spatangoid *Meoma ventricosa* (Fig.1j) (Chesher 1969), and in the clypeasteroid *Dendraster excentricus* (Fig.1g) as well as in *Strongylocentrotus purpuratus* (Fig.1d) (Fenner 1973). Ciliary currents on the exterior of the ampullae are arranged in parallel so that perivisceral fluid oxygenated at one ampullar surface does not cross another, an arrangement that maximizes the diffusion gradient across every ampulla (Fig.1d). The anatomical adaptations allowing countercurrent flows are also found in the flattened C1 tube feet of regular echinoids, and in the highly specialized respiratory tube feet in the irregular echinoids. The counter-current arrangement should allow a nearly complete exchange of gases first between the ambient medium and the fluid in the tube

Figure 1. Respiratory adaptations in the water vascular system of echinoids. Solid arrows represent currents external to, and broken arrows ciliary currents within, tube feet and ampullae. Drawings not to same scale. a) Longitudinal section of P1 tube foot of *Centrostephanus nitidus,* showing dual pores and central collagenous septum (After Smith 1978). b) Longitudinal section of P2 tube foot and ampulla of *Strongylocentrotus purpuratus;* note thicker wall relative to P1 tube foot, and counter-current arrangement of flows associated with ampulla (After Phelan 1967, and Smith 1978). c) Cross-section of P2 tube foot of *S.purpuratus* at level of septum; note separation of incurrent and excurrent channels (After Fenner 1973). d) Ampullae of *S.purpuratus,* viewed from inside test; note arrangement of internal and external currents (After Fenner 1973). e) Longitudinal section of C1 tube foot of *Eucidaris metularia;* note thin walls and separation of incurrent and excurrent channels by central convoluted area (After Smith 1978). f) As (e), sagittal section, showing increased surface area provided by extensions of convoluted area (After Smith 1978). g) Longitudinal section of respiratory tube foot and ampulla of *Dendraster excentricus;* note accessory pores in test, and counter-current arrangement of flows associated with tube foot and ampulla (After Fenner 1973, and Phelan 1977). h) Longitudinal section of respiratory tube foot and ampulla of *Echinocardium cordatum;* note relatively large pores in test, and multiple septa and passageways between incurrent and excurrent channels (After Smith 1980). i) As (h), sagittal section, showing increased surface area provided by lateral folding of tube foot, and flattening of ampulla (After Smith 1980). j) Portion of a respiratory (petaloid) ambulacrum of *Meoma ventricosa;* note pattern of water flow along ambulacrum and between respiratory tube feet, and counter-current flows associated with tube feet and ampullae (After Chesher 1969).

foot, and subsequently between ampullar fluid and the perivisceral coelom.

Adoption of an infaunal existence by irregular echinoids has required special respiratory adaptations. An adaptation to living within the substratum and a flattening of the test is the concentration of the respiratory ambulacra frequently in a petaloid arrangement, with their highly modified tube feet on the aboral surface, in closest proximity to the oxygen supply (Kier 1974). Those tube feet on the oral surface most removed from ambient oxygen but closest to the food-rich substratum became specialized for feeding. Still other tube feet were adapted for burrow maintenance and sensory reception (see Nichols 1959a). Perhaps as a consequence of such division of labor and the increase in efficiency of each type of tube foot for its particular task, the total number of tube feet in irregular echinoids is less than that in regular echinoids: the total number in a 'typical' regular echinoid of average size, *Strongylocentrotus droebachiensis,* is 1000 (Steen 1965); the number in average size spatangoids ranges from 212 to 385 (Nichols 1959a, p.357). Development of gas exchange capacity has been emphasized in these infaunal forms, since between 28 and 57 % of their total tube feet are respiratory. Despite their low absolute numbers, an increased oxygen delivery capability due to greater efficiency of these specialized respiratory tube feet might be inferred from physiological data (see section 5).

Smith (1980) discerned a morphological series of increasing respiratory effectiveness in certain tube feet of irregular echinoids, and much of the following discussion derives from his work. Most notable is a reduction in height and concurrent lateral extension of the tube foot. The development of 'sunken' petals in spatangoids increases their effective width, which in turn increases the possible width of the lamellar tube feet within them (Kier 1974). Although as Smith suggests, the need to protect such structures from abrasion by sediment in these burrowing echinoids may have been a selective pressure in reducing the height of the tube foot, hydrodynamic factors were also involved. For example, although the distance travelled by the fluid in a lamellar tube foot is somewhat less than that in a tall tube foot such as that in *Sclerodactyla briareus* or a P-type tube foot in a regular echinoid, the flow is unidirectional and parallel to the test over much of the length of the lamellar tube foot. This enables the establishment of an external counter-current and more efficient gas exchange, as discussed for ampullae.

Both respiratory tube feet and their ampullae are exceedingly thin-walled and consist only of a single layer of epithelial cells on either side of a very thin connective tissue layer, reducing diffusion distance. Muscularization is minimal. Mucus glands, so prevalent in other tube feet (e.g. Nichols 1959b, c, Smith 1978, 1980), appear to be absent from respiratory tube feet. This is important not only from the standpoint of division of labor among tube feet, but also the consideration that mucus presents a particular barrier to oxygen diffusion (Ultsch & Gros 1979).

Whereas all other tube feet in extant cassiduloids and spatangoids are served by single pores in the test, respiratory tube feet always have two pores (Kier 1962). This permits a one-way flow of fluid through the tube foot-ampullar system, within which fluid is also channelled by septa or strands of connective tissue. These septa additionally act as struts to strengthen and maintain the shape of the thin-walled gas exchangers, analogous to the pillar cells in fish gill lamellae (Fenner 1973). Particularly wide tube feet, such as those in the clypeasteroid *Dendraster,* are served by additional adradial pores (Durham 1966) through which incurrent fluid is more evenly distributed to individual channels in the thin central region of the tube foot (Fenner 1973, Phelan 1977). Very wide tube feet typically are skewed so as to be tallest adradially, where there is a large vertical passageway into the tube feet, which promotes distribution of incoming fluid among the central channels (Smith 1980).

An increased surface area of respiratory tube feet is provided by their flattened shape and by the folding or rippling of their walls to form sidebranches (Fenner 1973, Smith 1980). The latter adaptation is prevalent among the deep-burrowing spatangoids.

'Respiratory' tube feet are so designated by virtue of their anatomy, morphology and location relative to water currents (see section 3). Although a gas exchange function for tube feet with counter-current flow can be argued from first principles, none of the presumed respiratory specializations has been corroborated by direct physiological measurements (Fenner 1973). Nevertheless, the striking convergence in anatomy (counter-current flow; strengthening septa) and morphology (lamellar shape; arrangement in register in the petaloid ambulacra) between these podia and established respiratory structures, the secondary lamellae of fish gills, are compelling evidence for a functional convergence as well.

3 EXTERNAL CONVECTION MECHANISMS

Mechanisms for creating convection of the external medium are best developed in forms whose morphology or mode of life (especially the adoption of the infaunal habit) removes them from ready access to environmental oxygen. These include members of the asteroid families Astropectinidae, Goniopectinidae, Luidiidae, and Porcellanasteridae, as well as the irregular echinoids (orders Clypeasteroida and Spatangoida), most holothuroids, and many ophiuroids. Cloacal pumping is present in some extant crinoids, as it may have been in the extinct camerate crinoids (Haugh 1973) and in the extinct isorophinid and lebetodiscinid edrioasteroids (Bell 1977). Cloacal pumping in the latter would seem related to delivering oxygen to tissues within their subglobular, massively plated tests, rather than to low external availability of oxygen, since these suspension feeders likely occupied habitats where the water was well mixed.

Cilia provide the driving force for external convection in virtually all asteroids, a fact likely related both to their comparatively flattened body and to the occupation of easily disturbed, fine sediments by infaunal forms such as goniopectinids and porcellanasterids (Shick *et al.* 1981). In these families a heavily ciliated epithelium is elaborated into folds to create the cribriform organs in the interradial margins (Sladen 1883, Madsen 1961, Shick *et al.* 1981). This arrangement creates partially enclosed passages and increases the number of cilia per unit cross-sectional area within them set at right angles to the current. The cribriform organs thus act as 'turbines' and create a high pressure flow. High pressure may be necessary to maintain the multiple excurrent openings to the burrow from the arm tips which are excavated hydraulically: generally there is no physical contact between the arm tips and the sediment surface. Members of the Astropectinidae, which lack cribriform organs, typically assume a burrow posture which maintains contact between the arms and the sediment surface (Christensen 1970), although this may be related to a sensory function of the arm tips in this predatory group.

Although the cribriform organs are superficially similar to the pectinirhombs (the principal gas exchange organs) in extinct cystoids (Paul 1968, 1977), the cribriform organs create the respiratory current but are not the major sites of oxygen uptake: the diffusion distance from the cribriform lamellae to the coelom is rather large (ca. 0.5 to 1 mm: Fig.4 in Shick *et al.* 1981), so that the papulae of the already thin aboral surface likely are the principal route of gases to and from the perivisceral fluid. Stellate paxillae and sheets of mucus suspended from them protect the papulae from abrasion and fouling by sediment particles.

Rather than being a respiratory adaptation *per se*, the muscle-driven tidal irrigation of the papulae in the nidamental chamber of *Pteraster tesselatus* is a secondary feature necessitated by the mucoid defense mechanism in the supradorsal membrane of this asteroid (Nance & Braithwaite 1981), which occupies a habitat comparatively benign from a respiratory standpoint. The muscular fan spines adjacent to the incurrent respiratory pores in the ambulacra apparently do not create local water currents, but may serve to channel water into the pores or to keep them clear of obstructing particles (J.M.Nance, personal communication).

Morphological adaptations to promote gas exchange in the regular echinoids involve primarily the development of specialized respiratory tube feet rather than the enhancement of external convection, as this group relies entirely upon the general body surface ciliation unsupplemented by special current-augmenting devices. However, extension and waving of the tube feet under hypoxic conditions no doubt promotes gas exchange. The echinoids were rather late in developing the burrowing habit, apparently not doing so until the Lower Jurassic (180 million years ago) (Kier 1974, Paul 1977), and the subsequent explosive radiation of the irregular echinoids is associated with the adoption of an infaunal, deposit-feeding lifestyle (Paul 1977). It is in these infaunal irregular echinoids that the relationship between increasing separation from the ambient oxygen supply and enhanced potential for convection is manifest.

The spatangoid *Echinocardium cordatum* burrows to a depth of 15 cm in sandy sediments, and uses specialized dorsal tube feet to maintain a respiratory funnel to the surface (Nichols 1959a). On its dorsal surface this echinoid possesses a well-developed inner fasciole, composed of heavily ciliated clavulae, to augment water flow down the respiratory funnel into the burrow and along the respiratory ambulacra. The funnel shape of these ambulacra aids in channeling water along their length. Foster-Smith (1978) noted that *E.cordatum* generates high pressures compared with other animals employing ciliary pumps and that its pumping rate may also be substantial. He attributed this to the densely-ciliated fascioles, which extend the cilia away from the test and thereby increase the number set at right angles to the direction of water flow at the incurrent (dorsal) and excurrent (subanal) points of the burrow. The dorsal fasciole of *E.pennatifidum* occupies a smaller area. This is related to this species' usual occupation of coarser deposits in which water can readily be drawn through the interstices. Conversely, *Brissopsis lyrifera* inhabits thick blue mud and has a proportionately large dorsal fasciolar area (Nichols 1959a). *Lovenia elongata* possesses a dorsal fasciole but lacks funnel-building tube feet and seems restricted to shallow burrowing in clean sand (Ferber 1976). Thus, the absence of both a dorsal fasciole and of funnel-building tube feet in *Spatangus purpureus* is explained by this echinoid's burrowing in coarse shell gravel, again drawing its respiratory current not down a single narrow tube but between the large particles above the whole animal, and using the general ciliation of the body to create the flow (Nichols 1959a). The tube feet in its petaloid respiratory ambulacra are arranged at obtuse angles to the long axis of the gentle flow, which places the maximum number of respiratory surfaces in the oxygenated stream.

In more primitive irregular echinoids (the *Micraster* group of fossils), all species lacked dorsal fascioles and burrowing was restricted to comparatively shallow depths (Nichols 1959a). However, Nichols inferred enhanced external convection across the respiratory tube feet from the increased ornamentation (bumps and ridges), which presumably increased the area of ciliated cells, associated with the respiratory ambulacra. Similarly, the depression of the petals in extant spatangoids both increases the area available for cilia and aids in channeling water over the respiratory tube feet (Kier 1974).

Clearly, the nature of the sediment affects the depth of burrowing. *E.pennatifidum* burrows deeper in coarse shell gravel than in more compact sand (Nichols 1959a). Likewise, *Meoma ventricosa* does not burrow deeper than the length of its apical spines (which maintain a respiratory channel) in fine sand or silt, but is found in clean sand at depths of up to 10 cm (Chesher 1969). The cassiduloids *Apatopygus recens* (Higgins 1974) and *Cassidulus caribbearum* (Gladfelter 1978), also burrow up to 20 cm deep in coarse-grained, porous sand despite their lack of morphological adaptations for maintaining direct contact with the water column.

Irrigation rates are affected by the resistance of the sediment to water flow. *Echinocardium cordatum* maintains a higher flow rate in sand, where excurrent water can percolate through the sediment at the posterior of the individual, than in mud, where water must exit via abandoned respiratory funnels (Foster-Smith 1978). Chesher (1969) also noted that since the permeability of substrata decreases with increasing depth in them, disposing of water may present a greater problem than obtaining it, particularly when the apical spines maintain an incurrent opening. Thus, spatangoids have anal and subanal fascioles and associated tube feet which create drainage or sanitary tunnels posteriorly, allowing the removal of wastes and preventing the recycling of water across the test.

The success of *E.cordatum* in a variety of substrata involves the use of mucus to stabilize the incurrent channel, especially in sandy sediments which lack the cohesiveness of mud or gravel. This is effected by the funnel-building tube feet which plaster the walls of the respiratory funnel with mucus during their extension (Nichols 1959a). Similarly, *Moira atropos* burrows to a depth of 15 cm in soft mud and stabilizes its funnel with mucus (Chesher 1963).

The lunules in the sand dollar *Mellita quinquiesperforata* bring water up from the sand beneath the individual by the phenomenon of 'induced flow' when water moves across the dorsal surface (Alexander & Ghiold 1980). This process was discussed as a feeding adaptation, but a role in irrigating external tissues buried in the sand is also possible.

Gislén (1924) was unable to demonstrate ciliary currents on the external surfaces of holothuroids, and respiratory current-generating mechanisms in this group appear to be exclusively muscular. The mechanisms of the alternate filling and emptying of the respiratory trees (water lungs) were reviewed by Farmanfarmaian (1966), Binyon (1972), and Wolcott (1981), and will be treated only briefly here. The process of lung-filling in holothuroids is more complex than in mammals but similar to that in frogs since it involves the sequential creation of negative and positive pressures in two distinct stages. First, water is drawn into the cloaca, past the relaxed anal sphincter, by a negative pressure created by the dilation of the relaxed cloacal wall through the contraction of the suspensor muscles with their origins on the body wall. The flexible body wall does not collapse during this process because of the greater area for attachment of the suspensor muscles on it than on the cloaca, allowing a net force for cloacal expansion (Wolcott 1981). Local facultative changes in body wall rigidity may also be involved. Second, the anal sphincter closes, the suspensor muscles relax, and water is forced into the respiratory trees past their sphincters under pressure created by the contraction of the circular muscles of the cloaca. Unlike the situation in air-breathers, lung-filling in holothuroids typically involves a stepwise accumulation of water in the respiratory trees (Crozier 1916, Ogawa 1928, Pantin & Sawaya 1953, Newell & Courtney

1965, Robertson 1972, Brown & Shick 1979), likely necessitated by the greater viscosity and inertia of the respiratory medium and the highly branched and fragile nature of the water lung. Exhalation is effected by relaxing the sphincters and allowing the coelomic pressure, which generally exceeds 1 cm H_2O (Pantin & Sawaya 1953), to collapse the respiratory trees. The forceful spouting sometimes seen (Pearse 1908) implies an active component of exhalation as well.

The divergence between asteroids and echinoids, on one hand, and holothuroids on the other, in the mechanism of external convection, appears related to basic differences in their body plans and modes of life. Asteroids and echinoids are essentially epifaunal in design. The use of cilia in external convection in secondarily infaunal forms has required the development of structures (e.g. cribriform organs and fascioles) to augment water flow over comparatively short distances. The elongate shape and hydrostatic skeleton of holothuroids as adaptations for burrowing suggest that they are fundamentally infaunal (see Clark 1964), as does their bilaterality (also seen in burrowing echinoids; Kier 1974). Foster-Smith (1978) has commented on the inadequacy of cilia in providing sufficient respiratory flow in comparatively large, long, burrowing animals, and burrow irrigation in infaunal polychaetes is indeed primarily muscular. In the infaunal, vermiform *Leptosynapta inhaerens* (a member of the Apodida, which lack respiratory trees), burrow irrigation is again accomplished by peristaltic muscular activity of the very thin body wall (Mangum & Van Winkle 1973). The presence of respiratory trees is associated with a more massive body wall and greater internal complexity than is seen in the Apodida, since the gas exchange apparatus ramifies within the coelom and reduces diffusion distances to internal tissues. In conjunction with the flexible body wall, the direct delivery of oxygenated water by the respiratory trees may have enabled epifaunal holothuroids to exploit creviculous habitats inaccessible to sympatric echinoids with rigid tests.

Ophiuroids use both ciliary and muscular mechanisms to irrigate their bursae. Ciliary currents associated with the bursae were seen in virtually all of the species examined by Gislén (1924). Gislén also reported muscular activity of the disk pumping water into and forcefully out of these chambers in *Amphiura filiformis, A.chiajei,* and *Asteronyx loveni.* Pentreath (1971) described the arrangement of the musculature that would expand and contract the bursae, and noted that such contractions would not necessarily be visible externally.

4 THE ROLE OF BEHAVIOR IN EXTERNAL CONVECTION

The suspension-feeding lifestyle in many echinoderms assures their position in flowing water which normally will be well-oxygenated. This is particularly so in most crinoids, and in many ophiuroids and holothuroids. The epibenthic sand dollar *Dendraster excentricus* assumes a suspension-feeding posture (Timko 1976) that places its respiratory tube feet in the current. Reef-dwelling echinoderms in general may benefit from water movement in their habitat (Russo 1977), and even infaunal forms such as *Echinolampas crassa* and *Mellita quinquiesperforata* utilize subsurface flow caused by water moving across the substratum (Thum & Allen 1976, Alexander & Ghiold 1980).

The epifaunal *Holothuria forskali* crawls toward a source of oxygen when local P_{O_2} is low (Newell & Courtney 1965); the site of the oxygen sensor involved is unknown, but is probably different than that involved in the regulation of cloacal pumping. Under conditions of declining P_{O_2}, the epifaunal *Asterias forbesi* and *A.vulgaris* crawl upward, which in their natural habitat would likely expose them to better oxygenated conditions (Shick 1976).

In infaunal echinoids the attenuation of normal oxygen supply typically results in their failure to burrow or the emergence from the substratum if already buried. This has been known at least since Gislén's (1924) laboratory observations on *Brissopsis lyrifera, Echinocardium cordatum,* and *Spatangus purpureus. E.cordatum* will attempt to reopen an occluded respiratory funnel, but repeated funnel closure leads to the echinoid's complete emergence (Nichols 1959a). In a field study Chesher (1969) interpreted the nocturnal emergence of large specimens of *Meoma ventricosa* as a response to decreasing oxygen levels, and confirmed the emergence response of this species, and of *Moira atropos, Brissopsis atlantica* and *Brissus unicolor,* during hypoxia in aquaria. The failure of young *M.ventricosa* to emerge at night was attributed to its better-developed subanal fasciole and larger surface:volume ratio than in the adults. *Plagiobrissus grandis,* another spatangoid sympatric with *M.ventricosa,* does not emerge at night due to its more efficient burrow irrigation. Reduction of P_{O_2} to 50 % air saturation also triggers emergence in *Cassidulus caribbearum* (Gladfelter 1978). *Lovenia elongata* emerges from the sand at low tide and when the overlying water is deoxygenated; return of the tide and aeration of the water each resulted in burrowing by this spatangoid (Ferber & Lawrence 1976).

Laboratory observations on buried specimens of *Ctenodiscus crispatus* indicate that the highly distensible apical cone maintains the incurrent opening to the burrow (Shick 1976, Shick *et al.* 1981). The rapidity with which the mud star re-establishes the burrow opening if it is occluded suggests that the initial stimulus is tactile. Prolonged hypoxia alone does not reliably elicit the extension response, but acts synergistically with hydrogen sulfide to cause extension and peristalsis of the cone (Shick 1976). Christensen's (1970) observations on *Astropecten irregularis* suggest a similar role for its cone, and this is also likely the case for this structure in deep sea, infaunal porcellanasterid asteroids (Madsen 1961). Apical cones are not formed by buried specimens of *Luidia clathrata* (J.M.Lawrence, personal communication).

The infaunal holothuroid *Sclerodactyla briareus* gradually extends its posterior from the burrow and into the overlying water as P_{O_2} declines from air-saturation, so that at 40 mm Hg about one-third to one-half of its body surface is exposed (Brown & Shick 1979). This behavior is associated with the increasing importance of the numerous tube feet in maintaining oxygen uptake under progressive hypoxia, relative to uptake via the respiratory trees. At low P_{O_2} *Stichopus parvimensis* fully extends its tentacles and elongates its body, both of which increase the surface:volume ratio and promote oxygen uptake (Dimock 1977).

Woodley (1975) commented on the insufficiency of bursal pumping alone to irrigate the burrow in infaunal amphiurids, and greatly extended Des Arts' (1910) original observations of the mechanism of convection. Amphiurids irrigate their burrows by undulating one or more arms in the vertical plane, the waves passing along the arms from the sediment surface toward the disk. Excurrent water flows up the other limbs of the burrow in which the arms are motionless. Incurrent flow is enhanced by the posture of the tube feet which block gaps between the arm spines, and by the construction of 'respiratory fringes' composed of mucus and sediment particles which also clog the spines. The burrow itself is stabilized by mucus in some species.

Comparatively little is known of the ventilatory response to reduced P_{O_2}, primarily because of the difficulty of measuring water flow in most echinoderms. However, amphiurid ophiuroids increase the frequency of arm undulations from 15 to 100 per minute in stagnating water (Woodley 1975), which implies an increasing ventilatory volume.

5 FUNCTIONING OF THE GAS EXCHANGE SYSTEM

It has been assumed that the above anatomical, morphological, and behavioral adaptations affect respiratory performance, especially in burrowing echinoderms. This section treats the manifestations of these adaptations in the physiology of echinoderms, and in particular, the role of convection.

External convection has been most studied in the holothuroids having respiratory trees, largely because of the holothuroid's convenient design, if not its behavioral tractability (Brown & Shick 1979, Wolcott 1981). The temporal separation of inhalant and exhalant flows allows measurements comparable to those made on vertebrates, although the situation is complicated by the capacity of holothuroids to use both the actively-ventilated respiratory trees and the passively irrigated general body surface in gas exchange. In addition to the rate of oxygen uptake via both of these routes, one can measure the ventilation volume ($\dot{V}$w) and the percentage extraction of oxygen from the current ($\%Extr_w$), although the latter has rarely been done directly and indirect estimations have led to some inconsistencies. The contribution of oxygen consumption by the respiratory trees themselves to $\%Extr_w$ has not been assessed. Considering the

Table 2. External convection and oxygen uptake in echinoderms under normoxic conditions. Values marked by an asterisk (*) have been calculated from the other variables in the equation $\dot{V}w/\dot{V}_{O_2} \cdot \%Extr_W \cdot Ci_{O_2} = 1.0$. Terms are defined in the text. Ci_{O_2} has been calculated from Pi_{O_2} using standard tables. For unpublished data, values given are means ± 1 S.E. and N = number of specimens (with at least three determinations on each, in the case of $\%Extr_W$). $\dot{V}_{O_2}$ refers to oxygen uptake across the irrigated surface only: in the case of *C.crispatus* and the echinoids, this is the whole body surface; in *P.tesselatus* and the holothuroids, this is the nidamental chamber and the respiratory trees, respectively. For *S.purpureus*, the first value of Ppf_{O_2} is for a single specimen without gonads, while values in parentheses are for ripe specimens. Data for a given species may be compiled from several sources.

Species	Pi_{O_2} (mm Hg)	$\dot{V}_{O_2}$ ($ml \cdot g^{-1} \cdot h^{-1}$)	$\%Extr_W$	$\dot{V}w$ ($ml \cdot g^{-1} \cdot h^{-1}$)	$\dot{V}w/\dot{V}_{O_2}$ (l/ml)	Ppf_{O_2} (mm Hg)	$\%Ew_{O_2}$	Reference
Asteroidea								
Ctenodiscus crispatus (5°C; 1.5 g dry weight)	159	0.020	8.2	33.8*	1.69	135	54	Shick *et al.* 1981
Pteraster tesselatus (10-12°C; 300-400 g wet weight)	140-150	0.007	18	6	0.86	100	54	Johansen & Petersen 1971
Echinoidea								
Echinocardium cordatum (≃15°C; ≃30 g wet weight)	140-155	0.006	30*	3	0.50	–	–	Montuori 1913, Foster-Smith 1978
E.pennatifidum (15°C; 20 g wet weight)	147-155	0.006 ± 0.001 (N = 3)	23.6 ± 3.6 (N = 6)	4.7*	0.78	55 ± 8 (N = 5)	37	Shick, unpub.
Spatangus purpureus (15°C; 20-40 g wet weight)	147-155	0.005 ± 0.002 (N = 3)	29.4 ± 3.1 (N = 4)	3.2*	0.64	86 (57 ± 2) (N = 6)	69	Shick, unpub.
Holothuroidea								
Cucumaria frondosa (10°C; 10 g dry weight)	140-150	0.022	16	19	0.86	102	59	Brown & Shick 1979
Holothuria forskali (15-17°C; 10-15 g dry weight)	147-155	0.019	21.0 ± 0.4 (N = 3)	16*	0.84	88 ± 6 (N = 4)	52	Newell & Courtney 1965, Shick, unpub.
Sclerodactyla briareus (10°C; 5 g dry weight)	140-150	0.013	12	13	1.00	100	42	Brown & Shick 1979

comparatively high $\dot{V}_{O_2}$ of these tissues *in vitro* (Herrera & Plaza 1981), and the volume of respiratory water entering and leaving them (Table 2), such a contribution could be significant.

Directly-measured oxygen extraction ranges from 12 to about 30% of the available oxygen under air-saturated conditions (Table 2). The higher values of 49% (Newell & Courtney 1965) and 54% (Robertson 1972) were calculated from the ventilatory volume and the change in oxygen content of the medium, and assumed that water passed into and out of the respiratory trees only once, a criterion that is not likely to be met in closed-vessel experiments (Brown & Shick 1979). A simple calculation suggests that these estimated values are indeed too high. For example, if the incurrent water has a P_{O_2} of 140 to 150 mm Hg, and if $\%Extr_W = 50$, then the partial pressure of oxygen in exhaled water (Pe_{O_2}) would be 70 to 75 mm Hg. Thus, for oxygen to diffuse from the water in the respiratory trees into the perivisceral fluid, the P_{O_2} of the latter would have to be less than 70 mm Hg. This does not appear to be so: perivisceral P_{O_2} in *Holothuria forskali* is 88 mm Hg and is at least 70 mm Hg in laboratory measurements of other aspidochirotes (Table 1). Similarly, the *calculated* $\%Extr_W$ values in *Cucumaria frondosa* and *Sclerodactyla briareus* under air-saturated conditions are 35 and 42%, respectively, somewhat higher than the directly measured values of 16 and 12% (Brown & Shick 1979). If the calculated $\%Extr_W$ values were accurate, Pe_{O_2} would be 98 and 87 mm Hg, both of which values are lower than the measured Ppf_{O_2} levels of above 100 mm Hg (Table 1).

Under air-saturated conditions, both *C.frondosa* and *S.briareus* show a difference of 20 to 30 mm Hg ($\overline{\Delta P}_{O_2}$) between the exhaled water and the perivisceral fluid (calculated from data in Brown & Shick 1979). A similar level (31 mm Hg) occurs in *H.forskali* (Shick, unpublished) and in *Molpadia arenicola* (K.Johansen, personal communication). ($\overline{\Delta P}_{O_2}$ decreases at low P_{O_2} in all of these species.) Thus, the effectiveness of O_2 uptake ($\%Ew_{O_2}$), which is the ratio of the actual oxygen uptake to the maximum possible, given the diffusion gradient that exists, is also similar among the holothuroids (Table 2). The values for $\%Extr_W$, Ppf_{O_2}, $\%Ew_{O_2}$ (and $\overline{\Delta P}_{O_2} \simeq 35$ mm Hg) are comparable in the cushion star *Pteraster tesselatus* (Johansen & Petersen 1971; and Table 2), which also employs tidal ventilation of its primary gas exchange surface. These relationships are similar despite the presence of coelomic cell hemoglobins in the perivisceral fluids in some species (*C.frondosa* and *M.arenicola*) and its absence from others (*S.briareus, H.forskali* and *P.tesselatus*). Our knowledge of the oxygen-binding kinetics of holothuroid intracellular hemoglobins and localized P_{O_2} under physiological conditions remains insufficient to assess their quantitative importance, but their high oxygen affinities *in vitro* ($P_{50} \simeq$ 2-4 mm Hg; Manwell 1959, 1964, 1966, Terwilliger & Read 1972) suggest a minimal role in O_2 transport under normoxic conditions.

The original estimate of oxygen uptake via the respiratory trees was provided by Winterstein (1909), who capped the posterior end of *Holothuria tubulosa* and thus reduced $\dot{V}_{O_2}$ by 50 to 60%. Other experiments involving cloacal occlusion near air-saturation have yielded similar values: 53% in *Stichopus mollis* (Robertson 1972), and 47% in *Cucumaria frondosa* (calculated for an average animal of 10 g dry weight from data in Brown & Shick 1979). The smaller reduction of 38% in eviscerated specimens of *Stichopus parvimensis,* which lack respiratory trees (Dimock 1977), may result from an elevated oxygen demand by the remaining tissues during regeneration. The somewhat lower value of 29% in cloaca-occluded *Sclerodactyla briareus* (calculated for an average 5-g individual; Brown & Shick 1979) may be due to the ability of the remainder of the body surface to compensate for the loss of cloacal oxygen uptake, so that oxygen uptake via the body surface when the cloaca is blocked may be greater than uptake via that route when pumping can occur. That this is the case is indicated by simultaneous measurements of oxygen uptake via both routes: $\dot{V}_{O_2}$ (as percent of total $\dot{V}_{O_2}$) via the respiratory trees is then 59% in *S. briareus* (Brown & Shick 1979), in good agreement with the values of about 57% in the related *Thyonella gemmata* (calculated from data in Gentleman 1970), 60% in *Holothuria forskali* (Newell & Courtney 1965), and 63% in *C. frondosa* (Brown & Shick 1979). The closer correspondence of values obtained by simultaneous measurements (63%) and by cloacal occlusion (47%) in *C. frondosa* may indicate a lesser capacity for compensatory oxygen uptake by the general body surface in this species than in *S. briareus* (values of 59 and 29%). Body size is an additional factor influencing the morphological partitioning of oxygen uptake, since blocking the cloaca results in a greater reduction of $\dot{V}_{O_2}$ in large specimens of *C. frondosa* than in small ones (Brown & Shick 1979). Their small size (0.5 g dry weight) may account for the similarity of $\dot{V}_{O_2}$ (5% difference) in intact and eviscerated specimens of *Eupentacta quinquesemita* (Sabourin & Stickle 1981).

As external convection enhances oxygen delivery, particularly during diminished oxygen availability, the convective systems in true tidal water breathers (holothuroids and *Pteraster*) are responsive to changes in oxygenation of the medium, particularly in infaunal forms, which live in more stressful respiratory environments. In the epifaunal *C. frondosa* (studied by Brown & Shick 1979), the frequency of cloacal pumping cycles (several stepwise inhalations and a single, large exhalation constitute one cycle, roughly equivalent to 'tidal volume' in vertebrate physiological terminology) declines continuously with ambient P_{O_2}. The decline in frequency is sufficiently large that the hourly volume of water pumped ($\dot{V}$w) likewise declines, despite a 2.5-fold increase in tidal volume. The latter increase is in agreement with the increased amplitude of contraction of isolated cloacal circular muscle of *C. frondosa* under hypoxia (Lutz 1930). Associated with the decreased pumping frequency and increased tidal volume is a modest increase in $\%Extr_w$ and decrease in $\overline{\Delta}P_{O_2}$. This seems due to an increased residence time of water in the respiratory trees. The net effect on oxygen uptake via the respiratory trees is a transient increase as Pi_{O_2}

initially falls, a declining $\dot{V}_{O_2}$ immediately thereafter, and a stable $\dot{V}_{O_2}$ over the Pi_{O_2} interval of 50 to 90 mm Hg. Below 50 mm Hg, cloacal pumping becomes ineffective (the anus often remaining open), a large dead space of water not flushed out with each pumping cycle develops, and oxygen uptake via this route ceases.

In the usually infaunal *S.briareus,* the frequency of pumping cycles is maintained over a wider range of Pi_{O_2}, declines at intermediate P_{O_2}, and increases sharply just before ceasing at about 55 mm Hg (Brown & Shick 1979). After a slight initial decline, tidal volume stabilizes at intermediate Pi_{O_2}, and increases sharply just before ceasing. As in *C.frondosa,* $\%Extr_W$ increases as Pi_{O_2} declines and is greatest (31 %) at 70 mm Hg, where pumping frequency is at its minimum and tidal volume is near-maximal. The greater 'oxyregulatory' ability in *S. briareus* than in *C.frondosa* of the same size is due in part to the compensatory ventilatory response (increasing $\dot{V}w$ at low P_{O_2}) in the former, and probably reflects adaptation to periodically low oxygen conditions in its infaunal habitat.

The tidal convection system in *Pteraster tesselatus* (Johansen & Petersen 1971) behaves similarly to that in *S.briareus.* The frequency of 'breaths' increases by about 50 % as Pi_{O_2} declines from 150 to 50 mm Hg. These authors also inferred an increase in tidal volume. Since $\%Extr_W$ increases and $\overline{\Delta}P_{O_2}$ decreases over the same range, a circulatory response is implicated but would be exceedingly difficult to demonstrate. Like *C.frondosa, P.tesselatus* develops ventilatory arrythmia and a dead space at very low Pi_{O_2}.

Newell & Courtney (1965) measured $\dot{V}w$ and $\dot{V}_{O_2}$ simultaneously in *Holothuria forskali.* Their data for $\dot{V}w$ are more difficult to interpret because they varied the P_{O_2} of the water bathing the general body surface independently of that entering the cloaca. Filling of the respiratory trees appears more gradual and less regular than in other holothuroids, and the four-fold increase in $\dot{V}w$ between 95 and 65 % air saturation seems to arise from an increased pumping frequency, as tidal volume apparently declines. Since $\%Extr_W$ and $\dot{V}w$ are reported to be inversely related in this species, the response to hypoxia does not involve increased $\%Extr_W$ as in other echinoderm water breathers. However, $\%Extr_W$ was calculated indirectly and implies complete equilibration of perivisceral fluid and water pumped, which is unlikely.

Water flow has been directly measured in the spatangoid *Echinocardium cordatum.* A specimen of 5 cm test length (about 30 g live weight) moves an average of 90 ml H_2O per hour using its ciliary pump (Foster-Smith 1978). Knowing the $\dot{V}_{O_2}$ in this species and the oxygen content of the water pumped allows calculation of $\%Extr_W$ from the Fick principle (see equation in Table 2) as follows. A 27-gram individual has an oxygen uptake of 0.312 ml·h^{-1} at 24-25°C (Montuori 1913). Assuming a Q_{10} of 2, the oxygen uptake by an individual of this size would be 0.156 ml·h^{-1} at 15°C. Since 90 ml of water h^{-1} are pumped, 0.522 ml of oxygen would be available each hour (90 ml H_2O *times* 5.8 ml O_2 · liter H_2O^{-1} at 15°C). Dividing oxygen uptake by oxygen available gives a

%Extr$_W$ of 30 (Table 2). The agreement of this calculated value with those directly measured in *E.pennatifidum* and *Spatangus purpureus* (Table 2) increases confidence in the values of $\dot{V}w$ calculated for these species from measured *%Extr*$_W$, Ci_{O_2}, and $\dot{V}_{O_2}$. All of the spatangoids studied show similar values for these parameters.

Considering all echinoderms for which data are available (Table 2), there is a remarkable convergence in the convection requirement ($\dot{V}w/\dot{V}_{O_2}$), which derives in part from their generally low oxygen demand (see Lawrence & Lane 1982, for review and discussion of rates of O_2 uptake). The large convection requirement in holothuroids (average 0.9 l $H_2O \cdot$ ml O_2^{-1}) and *Pteraster*, relative to the spatangoids (average 0.64 l $\cdot$ ml^{-1}), probably relates to the muscular tidal irrigation of a voluminous internal chamber in the former as opposed to the unidirectional ciliary flow across a more compact respiratory surface in the latter. The higher value in *Ctenodiscus crispatus*, which employs ciliary currents, may be a consequence of the more diffuse flow in its burrow (with up to five excurrent channels), and of a smaller diffusion gradient owing to its high Ppf_{O_2}. Both factors would necessitate a greater flow of water. The convection requirement is somewhat low even in *C.crispatus* compared to that in other taxa using similar convection mechanisms. This indicates a primarily respiratory, as opposed to feeding, function of the current (Shick *et al.* 1981).

Modulation of water flow into the respiratory trees is involved in the relative independence of whole-organism $\dot{V}_{O_2}$ from ambient P_{O_2} in *C.frondosa,* since occlusion of the cloaca results in a large decrease in 'oxyregulatory' ability (Table 3). A smaller reduction is seen after evisceration in *Stichopus parvimensis* (Dimock 1977). The effect is also much less pronounced in *S.briareus* than in *C.frondosa* because the body surface compensates for loss of respiratory exchange via the respiratory trees and actually increases its contribution to whole-animal $\dot{V}_{O_2}$ as ambient P_{O_2} declines (Brown & Shick 1979).

In nature, this compensation is enabled by the emergence of the normally infaunal *S.briareus* from its burrow, thus directly exposing its numerous tube feet to the overlying water. Colacino's (1973) model for oxygen transport by the tube foot-ampullar system in this species used, among other parameters, the morphometry of the system, the total number of tube feet, the volume of fluid flow through them, the P_{O_2} gradient from environment to ampulla, and the characteristics of the cellular hemoglobin in the water vascular system. Despite the high oxygen affinity of the hemoglobin, Colacino concluded that 70 % of the O_2 transported by the system under air-saturated conditions was bound to hemoglobin. This surprising conclusion may arise from the difficulty in extrapolating from his data for kinetics of hemoglobin in solution to the actual kinetics of hemoglobin in red cells in the intact individual, and from the uncertainty of the P_{O_2} inside the ampulla. Moreover, since this holothuroid occupies a burrow which it does not irrigate, the tube feet are not generally exposed to air-saturated water, and the question of the role of the water vascular hemoglo-

Table 3. Indices of oxygen independence of $\dot{V}_{O_2}$ in echinoderms. B_2 is the quadratic coefficient in the polynomial regression $\dot{V}_{O_2} = B_0 + B_1(P_{O_2}) + B_2(P_{O_2})^2$, where more negative values indicate greater independence of $\dot{V}_{O_2}$ from ambient P_{O_2} (Mangum & Van Winkle 1973). P_c is the 'critical oxygen pressure' below which $\dot{V}_{O_2}$ becomes strongly dependent on P_{O_2}.

Species	B_2 (x 10^5)	P_c (mm Hg)	Reference
Asteroidea			
Asterias forbesi	–10.53	–	Mangum & Van Winkle 1973
Ctenodiscus crispatus	– 6.07	10-25	Shick 1976
Henricia sanguinolenta	– 8.39	–	Mangum & Van Winkle 1973
Luidia clathrata	– 0.90	30-40	Diehl *et al.* 1979
Pteraster tesselatus	–	<50	Johansen & Petersen 1971
Echinoidea			
Arbacia punctulata	– 4.73	–	Mangum & Van Winkle 1973
Echinarachnius parma	–13.92	–	Mangum & Van Winkle 1973
Echinocardium cordatum	– 9.83	–	Mangum & Van Winkle 1973
Strongylocentrotus purpuratus	–	60-70	Johansen & Vadas 1967
Holothuroidea			
Cucumaria frondosa			Brown & Shick 1979
(5 g individual)	– 6.81 (–3.31)[a]	–	
(10 g individual)	– 9.34 (–3.72)[a]	35	
Leptosynapta inhaerens	– 4.28	–	Mangum & Van Winkle 1973
Molpadia arenicola	–	50	K.Johansen, pers. comm.
Sclerodactyla briareus	– 7.30	–	Mangum & Van Winkle 1973
S.briareus (5 g individual)	– 9.89 (–6.72)[a]	40	Brown & Shick 1979
Stichopus parvimensis	– 6.89 (–6.22)[b]	50 (75)[b]	Dimock 1977 (B_2 recalculated)
Ophiuroidea			
Ophioderma brevispina	– 6.31	–	Mangum & Van Winkle 1973
Ophionereis reticulata	–	40	Petersen 1976
Ophiothrix quinquemaculata	–	30	Ott & Fedra 1977

a: Cloaca occluded; b: Eviscerated specimens

bin under high P_{O_2} conditions may not be relevant. The oxygen transporting role of hemoglobin in this holothuroid cannot be assessed by the classical technique of blocking it with carbon monoxide, since the partition coefficient for CO and O_2 is only 4:1 rather than the 100 to 200:1 for hemoglobin A and 2,000:1 for hemoglobin of the polychaete *Glycera dibranchiata* (C.P. Mangum, personal communication). Nevertheless, when using the value for mean maximum velocity rather than mean velocity of flow within the tube foot, the model predicts a net O_2 delivery via physical solution in water vascular fluid of 0.16 $ml \cdot h^{-1}$ in a 25-g (wet weight) individual. This is in excellent agreement with the measured rate of 0.15 $ml \cdot h^{-1}$ for the same size individual calculated from Mangum's (personal communication) weight-specific $\dot{V}_{O_2}$ of

0.015 ml·g^{-1}·h^{-1} and assuming a 41 % contribution of the body surface to total $\dot{V}_{O_2}$ (Brown & Shick 1979). The increasing role of the tube feet in oxygen transport as P_{O_2} declines is likely due to modulation of flow within this system and to an increasing involvement of its hemoglobin when P_{O_2} is very low. The mechanism for regulating flow is unknown.

Whereas the responsiveness of the tube foot-ampullar system is at least as important as that of cloacal pumping and increases in oxygen extraction in maintaining oxygen uptake during hypoxia in *S. briareus,* this cannot be the case in members of the infaunal order Molpadiida. Members of this group lack tube feet on the general body surface, so any systemic response to declining P_{O_2} must be mediated through their cloaca and respiratory trees. Few quantitative data on this group are available. Ogawa (1928) monitored cloacal pumping in *Paracaudina chilensis,* in which tidal volume appears to remain unchanged while pumping frequency declines with P_{O_2}. Times of confinement in the apparatus instead of actual P_{O_2} values are given, but effective ventilation apparently is maintained down to very low levels of oxygen. Nomura's (1926) assertion that $\dot{V}_{O_2}$ in this holothuroid 'is, at every instant, proportional to the oxygen tension of the medium' has often been repeated as an example of its rigid oxyconformity, although the relationship is actually hyperbolic and quite similar to that in the other holothuroids studied. It is rarely mentioned that his measurements were on a single group of ten individuals in the same vessel. Hiestand's (1940) report of oxygen-independence of $\dot{V}_{O_2}$ in *Sclerodactyla briareus* also involved groups of holothuroids. Both studies point out the difficulty in drawing conclusions from the data in the older literature.

The related *Molpadia arenicola* has a high degree of oxygen independence and a low P_C (Table 3) in burrowed specimens (K. Johansen, personal communication). Effective ventilation is maintained at lower P_{O_2} levels than in either *C. frondosa* or *S. briareus,* with both pumping frequency and tidal volume increasing under hypoxia, so that at 35 mm Hg $\dot{V}w$ is about twice that at normoxia. $\overline{\Delta P}_{O_2}$ declines over this range. $\%Extr_W$ at normoxia is about 30 %, rather high among directly-measured values in holothuroids. All of these relationships are probably related to the primary gas exchange role of the respiratory trees in this infaunal species that lacks a tube foot-ampullar system at the body surface.

It is significant that the apodid *Leptosynapta inhaerens,* which lacks both respiratory trees and tube feet (and hemoglobin), shows the highest oxygen dependence of $\dot{V}_{O_2}$ of any holothuroid studied (Table 3). Adoption of the burrowing lifestyle in this species involved extreme vermiformicity, attenuation of body wall thickness, and a peristaltic irrigation of the burrow, and the species' response to declining P_{O_2} is very much like that of burrowing polychaetes (Mangum & Van Winkle 1973). Absolute rates of oxygen uptake by *L. inhaerens* are similar to those in other holothuroids (Webster 1975, Lawrence & Lane 1982, C.P. Mangum, personal communication), although this has been achieved at the expense of body wall thickness and body size, and the effects of this on organic productivity are unknown. Comparable data on $\dot{V}_{O_2}$ and its dependence on P_{O_2} in large, epibenthic, reef-dwelling synaptids could be very informative, especially since body size has an effect on oxyregulatory ability (Table 3), and since the relative importance of the respiratory trees in oxyregulation increases with body size in *C.frondosa* (Brown & Shick 1979).

Within-class correlations of oxygen independence of $\dot{V}_{O_2}$ and the development of convection mechanisms and specialized gas exchange structures elsewhere in the Echinodermata meet with mixed success. Within the echinoids, the more specialized burrowing forms (*Echinocardium cordatum* and *Echinarachnius parma*) indeed are more oxyregulatory than the regular, epifaunal species *Arbacia punctulata* (Table 3). No pattern is seen in the asteroids, although the data of Johansen & Petersen (1971) for the apparently strongly-regulating *Pteraster tesselatus* do not allow calculation of an actual index of oxyregulation. The lack of oxyregulation indicated for the burrowing *Luidia clathrata* may be more apparent than real, as the $\dot{V}_{O_2} - P_{O_2}$ relationship is clearly multiphasic and poorly described by the polynomial model (Diehl *et al.* 1979).

The topological distribution and proportion of total biomass of the most actively respiring tissues may also be involved in the relationship between P_{O_2} and $\dot{V}_{O_2}$. Most of the total oxygen uptake in intact echinoids and asteroids is by the body wall and tube feet (see Giese 1976). Rather than representing simply the degree to which respiring tissues are insulated from their external source of oxygen (as suggested by Mangum & Van Winkle 1973), the quadratic term B_2 may reflect the fact that the preponderant contributors to total $\dot{V}_{O_2}$ are the superficial tissues which have ready access to oxygen and which are only slightly affected by a moderate decline in ambient P_{O_2}. This is especially the case in echinoids, and the ciliary enhancement of flow across the test may account for the dichotomy between the burrowing forms and *Arbacia punctulata* in their oxygen dependence. Whole-organism $\dot{V}_{O_2}$ in *Strongylocentrotus droebachiensis* is quite low, independent of P_{O_2}, and unaffected by covering most of the tube feet, at 6°C but not at 16°C (Steen 1965). These results are consistent with the test's being the major site of oxygen consumption. The reduction of metabolic demands at low temperature probably render the already low $\dot{V}_{O_2}$ by other body components undetectable, leaving only the test's contribution to be measured as 'whole-organism' $\dot{V}_{O_2}$.

Although it has been suggested that Ppf_{O_2} is unchanged or even increases as Pi_{O_2} initially declines (Newell & Courtney 1965, Johansen & Vadas 1967), this is not so in most echinoderms. The data in Figure 2 show that, with few exceptions, Ppf_{O_2} falls in more-or-less strict accordance with Pi_{O_2}, at least down to the intermediate levels of Pi_{O_2} where most observations ceased. The elevated Ppf_{O_2} at very low levels of Pi_{O_2} in *S.briareus* likely relates to an increasing role of its high affinity hemoglobin. An oxygen storage role of hemoglobin could

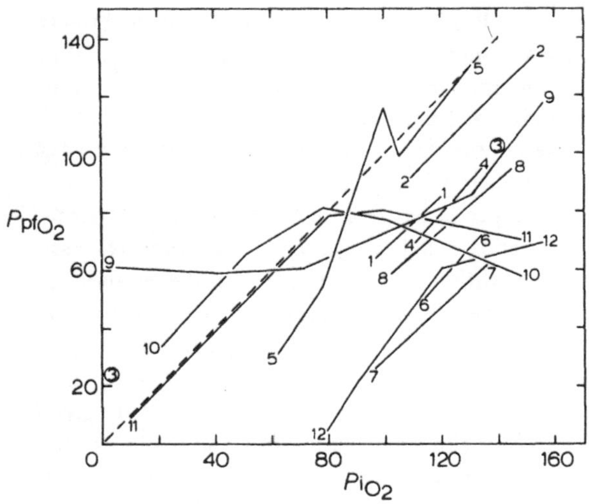

Figure 2. Effect of declining Pi_{O_2} on perivisceral fluid P_{O_2} in echinoderms. Dashed line represents the oxygen isobar. 1) *Clypeaster subdepressus* (Petersen, personal communication); 2) *Ctenodiscus crispatus* (Shick, unpublished); 3) *Cucumaria frondosa* (Brown & Shick 1979); 4) *Encope emarginata* (Petersen, personal communication); 5) *Holothuria forskali* (Newell & Courtney 1965); 6) *Holothuria grisea* (Petersen, personal communication); 7) *Luidia clathrata* (Petersen, personal communication); 8) *Pteraster tesselatus* (Johansen & Petersen 1971); 9) *Sclerodactyla briareus* (Mangum & Van Winkle 1973); 10) *Strongylocentrotus droebachiensis* (Johansen & Vadas 1967); 11) *Strongylocentrotus purpuratus* (Johansen & Vadas 1967); 12) *Strongylocentrotus purpuratus* (Webster & Giese 1975).

be considerable, especially in the burrowing molpadiids, in which perivisceral fluid represents about 30% of total body weight and whose oxygen concentration is seven times that of seawater (perivisceral O_2 concentration in *Paracaudina chilensis* is about 4 vols%; Kawamoto 1928).

Replenishment of depleted coelomic oxygen stores may account for some of the elevated $\dot{V}_{O_2}$ (oxygen debt phenomenon) following hypoxic exposure (Newell & Courtney 1965, Brown & Shick 1979). Following 12 h exposure to hypoxia, the Ppf_{O_2} in *Asterias vulgaris* increased rapidly during 1 h of recovery at normoxia, and reoxygenation of the perivisceral fluid accounted for about 10% of total $\dot{V}_{O_2}$ (Shick, unpublished). Reoxygenation of deoxyhemoglobin could produce an even larger increase in $\dot{V}_{O_2}$ in holothuroids, where an additional possibility is that the increased post-hypoxic $\dot{V}_{O_2}$ reflects the muscular activity involved in flushing acidic metabolites from the perivisceral fluid. Isolated longitudinal muscles of *S. briareus* release lactate to the external medium and do not show an oxygen debt (Ellington & Hammen 1977), while intact individuals do have an elevated $\dot{V}_{O_2}$ (Mangum & Van Winkle 1973, Brown & Shick 1979). The development of an oxygen debt in *S. briareus*, whether metabolic or systemic, depends on the extent of exposure to anoxia, since during short anoxic excursions the holothuroid uses its coelomic oxygen reserves and does not have an elevated post-hypoxic $\dot{V}_{O_2}$ (Brown & Shick 1979). In contrast, *C.frondosa* depletes its internal oxygen stores more quickly under progressive hypoxia (Fig.2) and shows an oxygen debt sooner. This reflects adaptations of these species to different frequencies of exposure to hypoxia in their natural habitats.

The metabolic and energetic significance of the oxygen debt response in most marine invertebrates is not well understood, in part because of the lack of a clear relationship between the amounts of various end-products other than lactate accumulated and the magnitude of the post-anoxic $\dot{V}_{O_2}$ increase. In holothuroids, which produce lactate as the principal anaerobic end-product (see Ellington 1982, for review), biochemical analysis would be facilitated and allow concurrent measurement of the above parameters, as well as the ventilatory response. Direct metabolic calorimetry (see section 6) in conjunction with $\dot{V}_{O_2}$ measurements would distinguish between the passive reoxygenation of the perivisceral fluid and a true increase in energy demand (heat production) during the recovery period.

The question remains whether the development of specialized gas exchange structures and convection mechanisms actually enhance oxygen delivery and affect the respiratory gas levels in the perivisceral fluid under ambient conditions. A comparison of the available data (Table 1) reveals that $P\text{pf}_{O_2}$ is highest overall in the holothuroids, which is certainly in accord with their possession of a mechanism to deliver O_2 directly throughout their internal fluids. Further, the highest value for $P\text{pf}_{O_2}$ is in *Sclerodactyla briareus,* which has an effective dual respiratory system involving the respiratory trees and the numerous tube feet. The lowest values for holothuroids are found in the aspidochirotes *Holothuria grisea* and *Isostichopus badionotus,* which have thick body walls and comparatively few and locally restricted tube feet. The few data on $P\text{pf}_{CO_2}$ available (Table 1) again support the *a priori* expectation: values are highest in the echinoids, intermediate in the asteroids, and lowest in the holothuroids.

Within the Asteroidea, it is probably significant that the highest $P\text{pf}_{O_2}$ is in *Ctenodiscus crispatus* which has an effective system for external convection, a very thin aboral integument, and a muscular apical cone to mix the internal fluids. A high $P\text{pf}_{O_2}$ also occurs in *Pteraster tesselatus* which likewise has an effective system for convection of the medium, but this generalization must be tempered by the fact that $P\text{pf}_{O_2}$ is just as high in reproductively spent specimens of *Asterias vulgaris* and *Marthasterias glacialis,* asteroids without remarkable respiratory adaptations.

None of the purported structural respiratory adaptations in echinoids has been corroborated by direct physiological measurements. Using the relatively crude, indirect physiological estimator of $P\text{pf}_{O_2}$, no clear pattern emerges and the values span a very wide range. Although quite high values occur in *Clypeas-*

ter subdepressus, Encope emarginata and *Spatangus purpureus* (which have specialized respiratory tube feet and effective convection), the values are equally high in *Paracentrotus lividus* and *Strongylocentrotus droebachiensis,* which have neither (see Appendix in Smith 1978). However, it must be remembered that the high $P\text{pf}_{O_2}$ in the irregular echinoids is achieved despite their smaller number of tube feet, implying an increased efficiency of the respiratory tube feet.

This does not mean that convection is without effect on $P\text{pf}_{O_2}$ and $P\text{pf}_{CO_2}$. J.A.Petersen (personal communication) showed that stirring the external medium increases $P\text{pf}_{O_2}$ and decreases $P\text{pf}_{CO_2}$ beyond the unstirred condition in *Echinometra lacunter* (Table 1). Johansen & Petersen (1971) reported that $P\text{pf}_{CO_2}$ in *Pteraster tesselatus* increases when ventilation is impaired at very low $P\text{i}_{O_2}$ or high temperature. Houlihan & Duthie (1981) found that $P\text{pf}_{O_2}$ in *Paracentrotus lividus* is positively correlated with the degree of exposure of the body from its rock burrow to the surrounding seawater in nature. Likewise, $P\text{pf}_{O_2}$ is greater and $P\text{pf}_{CO_2}$ less in *Astropecten brasiliensis* when the sea star is unburied than when buried (J.A.Petersen, personal communication).

The variability of $P\text{pf}_{O_2}$ among echinoids is related to additional intrinsic factors such as the thickness of the test, the body component accounting for most of the total $\dot{V}_{O_2}$ in this class (Giese 1976). Accordingly, Webster (1975) attributed the high internal P_{O_2} in *Allocentrotus fragilis* to its very thin test, which consumes proportionately less oxygen, leaving more of it at the body surface for transport to the perivisceral fluid. An additional explanation is that the thin test may enable oxygen entry directly across itself, rather than being restricted to entry via the tube feet as in species with thicker tests (see Farmanfarmaian 1966). This explanation may also apply to the spatangoids, which have thin, fragile tests, low $\dot{V}_{O_2}$s, and which have $P\text{pf}_{O_2}$ values as high or higher than other echinoids, despite their reduction in the number of tube feet. Nevertheless, the high internal P_{O_2} in *Clypeaster subdepressus* and *Encope emarginata* which have more massive tests, does imply an increased efficiency of their respiratory tube feet and not just a passive consequence of test thickness. The increased capacity for ciliary movement of water across the test in irregular echinoids itself should contribute to an enhanced delivery of oxygen to the perivisceral fluid. Finally, the presence of peristomial gills in regular echinoids affects local (peripharyngeal coelom) P_{O_2}, as discussed earlier.

A last important point is that the P_{O_2} in the perivisceral fluid represents the balance between oxygen delivery and oxygen consumption by the body wall and by internal tissues. Giese and his coworkers (Giese *et al.* 1966, Belman & Giese 1974, Webster & Giese 1975, Giese 1976) emphasized the importance of the echinoid body wall and relative unimportance of internal tissues in determining whole-organism $\dot{V}_{O_2}$ (values of approximately 90 and 10 % of total $\dot{V}_{O_2}$, respectively). This has been attributed to the exposure of the body wall to water with the highest P_{O_2}, resulting both in its high $\dot{V}_{O_2}$ and in a low P_{O_2}

in body fluids, which in turn causes a low $\dot{V}_{O_2}$ in internal tissues. This conclusion is justified by their data (all body components have similar rates of oxygen uptake when isolated in shaking respirometers under air-saturation), but the possibility (first suggested by Giese *et al.* 1966) that the $\dot{V}_{O_2}$ of gonad *in vivo* is not simply a consequence of low Ppf_{O_2} but actually contributes to coelomic oxygen depletion has not heretofore been demonstrated. The data in Table 1 for reproductively ripe and spent specimens of *Echinus esculentus* (9 to 11 cm in diameter) and *Spatangus purpureus* (5.5 to 7 cm test length) reveal a significantly lower Ppf_{O_2} in ripe individuals. Thus, oxygen delivery does not keep pace with an elevated internal oxygen demand caused by an increase in organ biomass. The range of Ppf_{O_2} values determined for ripe and spent similar-sized specimens of the same species at the same time is sufficiently great to obscure any differences attributable to structural or physiological adaptations, and emphasizes the need to specify reproductive condition in such studies.

That oxygen uptake by ripe ovaries is a major contributor to this situation can also be shown as follows. Allowing for slight differences in the volume of perivisceral fluid, the lower Ppf_{O_2} in ripe female *Echinus esculentus* relative to spent individuals (Table 1) is equivalent to a steady-state difference in oxygen content of about 0.88 ml O_2 per individual. This assumes that the oxygen capacitance (β_{O_2}) of the perivisceral fluid, which has no respiratory pigment, is the same as that of seawater, an assumption supported by the data in Houlihan & Duthie (1981). At a P_{O_2} of 30-45 mm Hg and 15°C, the total volume of oxygen consumed by intact ovaries of *E. esculentus* is 0.39 ml $O_2 \cdot h^{-1}$ (Shick, unpublished). Dividing 0.39 ml O_2 by the steady-state difference of 0.88 ml O_2 reveals that about half (45 %) of the difference in the oxygen content of the perivisceral fluid of spent and ripe females is accounted for by uptake by the ovaries.

6 SIMULTANEOUS AEROBIC AND ANAEROBIC METABOLISM

Although echinoderms are basically aerobic animals with only modest tolerance of anoxia (Theede *et al.* 1969, Ellington 1975, Shick 1976, Diehl *et al.* 1979), the metabolism in certain of their organs may be largely anaerobic, even under normoxic external conditions. Ellington (1982) has referred to this seemingly paradoxical situation as 'the problem of internal tissues', and the suggested evidence for it is indirect but manifold:

1. Low Ppf_{O_2} (Johansen & Vadas 1967, Webster & Giese 1975, Houlihan & Duthie 1981);
2. Low tissue P_{O_2} (Webster & Giese 1975);
3. Low capacity for aerobic metabolism in internal tissues (Mattisson 1959, Giese *et al.* 1966, Ellington & Hammen 1977, Hill *et al.* 1978);
4. Unbalanced energy budgets (Miller & Mann 1973, Greenwood 1980);

Table 4. Rates of heat production ($\dot{Q}_H$) and oxygen uptake ($\dot{V}_{O_2}$) at 15°C by various echinoderm tissues under anoxic, normoxic ($\simeq Ppf_{O_2}$), and hyperoxic (air- and O_2-saturation) conditions. Brackets indicate parallel experiments conducted on tissues from the same individuals. $\dot{Q}_H$ by tissues incubated in filtered seawater of various degrees of initial oxygenation was measured in a double-twin heat-flow calorimeter (Pamatmat 1979; Shick 1981). $\dot{V}_{O_2}$ of similarly-treated tissues from the same individuals was measured in parallel by discontinuous determinations of P_{O_2} in aliquots of the incubation water. $\dot{Q}_{OX}$ is the estimated heat production calculated from $\dot{V}_{O_2}$ using the oxycaloric coefficient of 20.3 J · ml O_2^{-1}. All rates are expressed on a wet-weight basis (Shick, unpublished data).

Tissue	Number of specimens	Number of determinations $\dot{V}_{O_2}$	Number of determinations $\dot{Q}_H$	Ambient P_{O_2} (mm Hg) $\dot{V}_{O_2}$	Ambient P_{O_2} (mm Hg) $\dot{Q}_H$	$\dot{V}_{O_2}$ (μl $g^{-1} h^{-1}$)	$\dot{Q}_H$ (J $g^{-1} h^{-1}$)	$\dot{Q}_{OX}$ (J $g^{-1} h^{-1}$)	$\Delta\dot{Q}_H$	$\Delta\dot{Q}_{OX}$	Anoxic $\dot{Q}_H$ / Normoxic $\dot{Q}_H$
Echinus esculentus Ovary	3	–	3	–	0	–	0.087 ± 0.026	–	0.039	0.045	0.69
	3	3	3	30-40	40-50	2.2 ± 0.26	0.126 ± 0.027	0.045			
Ovary	1	3	1	150	155	6.8 ± 1.0	0.167	0.138	0.140	0.124	
	1	3	1	650	600	12.9 ± 2.0	0.307	0.262			
Asterias rubens	1	–	1	–	0	–	0.172	–			
Pyloric caecum	1	–	1	–	71-96	–	0.253				0.68
Holothuria forskali	1	–	1	–	0	–	0.079	–			
Longitudinal muscle	1	–	1	–	86-100	–	0.097	–			0.81

5. Low pH in the perivisceral fluid (Mangum & Shick 1972, Houlihan & Duthie 1981).

Regarding the first point, Ppf_{O_2} is not inordinately low compared with P_{O_2} of body fluids in other marine invertebrates (*cf.* Mangum 1973), although the low value in ripe echinoids (in part caused by ovary $\dot{V}_{O_2}$) does depress tissue $\dot{V}_{O_2}$ below that under air-saturated conditions (Giese 1976; and Table 4). Nevertheless, the prevailing Ppf_{O_2} does not appear to impose a strict upper limit on tissue $\dot{V}_{O_2}$ *in vivo,* since Lilly (1979) has shown a 'specific dynamic effect' (elevated $\dot{V}_{O_2}$) after feeding in *Strongylocentrotus droebachiensis* and *Tripneustes ventricosus.* This is likely associated with the gut, which tends to have a higher $\dot{V}_{O_2}$ than most other tissues under similar conditions (Moore & McPherson 1965, Giese *et al.* 1966, Percy 1974).

Tissue (ovary) P_{O_2} is indeed low, ranging from 0 to 10 mm Hg in *Strongylocentrotus purpuratus* (Webster & Giese 1975); whether these values are for deep and superficial tissues, respectively, is not known, and the higher value is similar to that prevailing in a number of aerobic mammalian tissues (see Jöbsis 1977). In any case, the finding of anoxic conditions led Webster & Giese (1975) to their initial proposal of a largely anaerobic metabolism in internal tissues. Because echinoids lack a vascular system for the internal transport of oxygen, the low ovary P_{O_2} is more a consequence of lack of perfusion of this massive organ than of a sparse oxygen supply in the fluid bathing it. Incubation of ovaries of *Echinus esculentus* under hyperoxic conditions increases their $\dot{V}_{O_2}$ (Table 4), confirming that they are diffusion-limited under normoxia. A similar increase in $\dot{V}_{O_2}$ was seen in longitudinal muscle of *Parastichopus tremulus* under hyperbaric oxygen *in vitro* (Mattisson 1959). Studies on whole echinoderms also demonstrated increased $\dot{V}_{O_2}$ under hyperoxia (Maloeuf 1937, Johansen & Vadas 1967, Webster & Giese 1975, Petersen 1976), and despite the valid *caveat* that some of this could be due to increased oxygenation of the voluminous body fluids (Johansen & Vadas 1967), calculations from the data in Webster & Giese (1975) show that only 5 to 10 % of the increased oxygen uptake in *S.purpuratus* is so explained even if the perivisceral fluid equilibrated fully with the ambient medium. Presumably, therefore, internal tissue oxygenation and $\dot{V}_{O_2}$ also increase under these conditions.

A low capacity for aerobic metabolism in echinoderm tissues is indicated not only by their low rates of oxygen consumption (Mattisson 1959, Giese *et al.* 1966, Ellington & Hammen 1977, Lawrence & Lane 1982, Table 4, present paper) compared to other marine invertebrates, probably due to the relative lack of muscular development in the former (Lawrence & Lane 1982), but also by biochemical, enzymic, and ultrastructural evidence. These studies were summarized and discussed by Ellington (1982), who concluded a low energy demand and possible anaerobic metabolism in internal tissues.

An imbalance in energy budgets determined for echinoids may also indicate a contribution of anaerobic metabolism, which is never measured in such

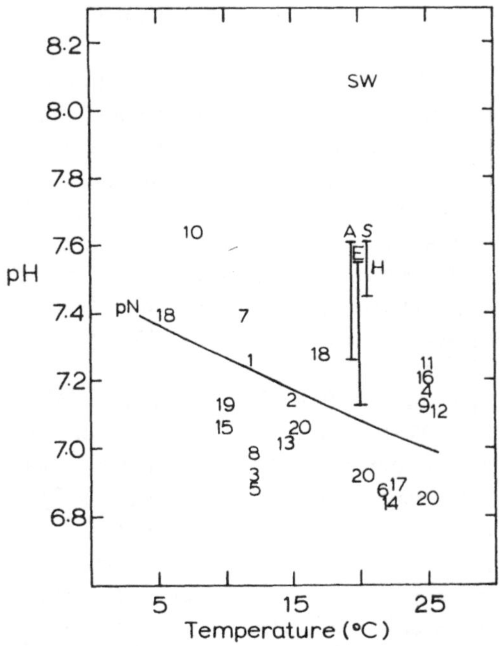

Figure 3. Body fluid pH in echinoderms. All samples (except those in Shumway 1977) were taken anaerobically, and all measurements were made at the temperature at which the animals were living in the field or were acclimated to in the laboratory. Vertical bars represent pH ranges predicted using maximum and minimum literature values for C_{CO_2} and P_{CO_2} (see text for discussion); A: Asteroidea; E: Echinoidea; S: *Strongylocentrotus droebachiensis* (using 4.15 and 6.00 mM C_{CO_2}, and P_{CO_2} = 2.8 mm Hg, the highest reported value for any regular echinoid at normoxia); H: Holothuroidea. SW represents the pH of standard seawater with C_{CO_2} = 2.2 mM and P_{CO_2} = 0.024 mm Hg. pN is the neutral point of pure water. Asteroidea: 1) *Asterias rubens* (Shumway 1977); 2) *Asterias vulgaris* (Shick & Dykens, unpublished); 3) *Astropecten irregularis* (Shumway 1977); 4) *Echinaster sepositus* (Houlihan & Duthie 1981); 5) *Henricia sanguinolenta* (Shumway 1977); 6) *Pisaster ochraceus* (Punzo 1977); 7) *Pteraster tesselatus* (Johansen & Petersen 1971); 8) *Solaster papposus* (Shumway 1977). Echinoidea: 9) *Arbacia lixula* (Houlihan & Duthie 1981); 10) *Echinus esculentus* (Houlihan, personal communication); 11) *Paracentrotus lividus* (Houlihan & Duthie 1981); 12) *Sphaerechinus granularis* (Houlihan & Duthie 1981); 13) *Strongylocentrotus droebachiensis* (Shick & Dykens, unpublished); 14) *Strongylocentrotus purpuratus* (Punzo 1977). Holothuroidea: 15) *Cucumaria frondosa* (Brown & Shick 1979); 16) *Holothuria polii* (Houlihan & Duthie 1981); 17) *Sclerodactyla briareus* (Mangum & Shick 1972); 18) *Sclerodactyla briareus* (Mangum 1973); 19) *Sclerodactyla briareus* (Brown & Shick 1979); 20) *Thyonella gemmata* (Gentleman 1970).

studies. In Miller & Mann's (1973) study of *S.droebachiensis,* an average of 42% of the ingested energy was unaccounted for in the balanced energy equation, where consumption, growth, defecation, and respiration (as $\dot{V}_{O_2}$) were measured. Greenwood (1980) reported discrepancies of 24 and 53 % in two populations of *Parechinus angulosus.* All of these workers validly invoked the unmeasured loss of dissolved organic matter by well-fed urchins (see Field 1972) as an explanation for the imbalance, but losses via ammonia excretion and via anaerobic fermentation are also possible contributors. Presumably the discrepancy would be greatest in ripe specimens with massive, anoxic gonads.

The pH of the body fluids of echinoderms (Fig.3) is generally lower than that in other marine invertebrates (Mangum & Shick 1972, Howell *et al.* 1973, Mangum 1973, Punzo 1977, Houlihan & Duthie 1981). This could be due to the release from the tissues of acidic organic metabolites (Mangum 1973, Brown & Shick 1979, Houlihan & Duthie 1981). Ellington & Hammen (1977) reported the presence of lactate in *Sclerodactyla briareus* perivisceral fluid. Alternatively, Farmanfarmaian (1966) attributed the low pH in echinoderm body fluids to the retention of respiratory CO_2. Unfortunately, there have been no studies of rates of carbon dioxide release in echinoderms.

The problem of carbon dioxide and perivisceral fluid pH warrants further attention, since the question of whether the relative acidosis of these fluids is respiratory or metabolic bears not only on the problem of aerobic *vs.* anaerobic metabolism, but also on that of acid-base balance and carbon dioxide excretion. The only direct study of the buffering capacity of fluids is that of Sarch (1931) who concluded that echinoderm body fluids, like seawater, are buffered primarily by the carbon dioxide-bicarbonate system, but a relationship among these parameters and pH cannot be determined from his data, and P_{CO_2} and temperature were not given. In the older literature (Collip 1920, Okazaki & Koizumi 1928, Sarch 1931, Cole 1940) there are values for the carbon dioxide concentration (C_{CO_2}) of perivisceral fluid, although the temperature and salinity are not specified (the latter is likely 30 ‰S or greater). The data, which must be viewed with caution, average 2.3 mM total carbon dioxide in asteroids, 2.1 mM in holothuroids, and 2.2 mM in most echinoids, although the average value in *S.droebachiensis* is 5.1 mM (4.2 mM reported by Collip 1920; 3.8 mM by Sarch 1931; and 6.0 mM by Cole 1940). These values are very low among aquatic animals, in which the range is 3 to 15 mM (Cameron 1979), and agree reasonably well with the more recent value of 2.8 mM C_{CO_2} in *Asterias rubens* (Binyon 1962). (Some of Cole's 1940 values indicate no difference between body fluids and seawater in C_{CO_2}, but his seawater samples were stored before analysis, and oxidation of dissolved organic matter in them may have raised their CO_2 content.)

Like C_{CO_2}, P_{CO_2} is low in echinoderm fluids (Table 1), the values in other marine animals ranging up to about 5 mm Hg (Dejours 1975, Cameron 1979). Thus, the low pH in echinoderm body fluids exists despite low values for C_{CO_2} and P_{CO_2}, and the extent that the pH is due to carbon dioxide is unknown.

As a first approximation, the maximal and minimal values from the literature for these parameters have been used in the standard equations for seawater (Skirrow 1975) to calculate their effect on pH at 20°C, a representative temperature of measurement. Such calculations represent a worst-case situation in that they assume that the body fluids are no better buffered than seawater, which is unlikely given their protein content (Giese 1966). Most of the measured pH values in a variety of echinoderms fall at or below the lower limits of the predicted ranges (Fig.3), indicating that low body fluid pH is due to factors in addition to CO_2. These factors may be the continuous presence of weak acids such as lactate, and strong ion difference (Stewart 1978), in the body fluids. A systematic study of acid-base balance in echinoderms has not been made. Such a study would be especially interesting in the holothuroids, which show the expected inverse relationship between body fluid pH and temperature yet which have particularly acidic body fluids (*S.briareus* and *T.gemmata,* Fig.3). The effects of hypoxia and ventilation on short-term acid-base variations measured by changes in P_{CO_2}, total CO_2 and bicarbonate, and lactate would be a fruitful area of research.

Direct calorimetric measurements of heat production ($\dot{Q}_H$) by various echinoderm tissues *in vitro* indicate a large anaerobic contribution to total metabolic rate, as anoxic $\dot{Q}_H$ in ovaries of *Echinus esculentus* and in pyloric caeca of *Asterias rubens* was 68-69 % of the $\dot{Q}_H$ under normoxic physiological conditions (Table 4). Incubation of ovaries under hyperoxia increased not only their $\dot{V}_{O_2}$ but also $\dot{Q}_H$, confirming that aerobic metabolism in this tissue is normally diffusion-limited.

Results on longitudinal muscle of *Holothuria forskali* were qualitatively similar, but indicate an even larger anaerobic component (81 % of normoxic $\dot{Q}_H$) despite the higher prevailing normoxic P_{O_2} (Table 4). This suggests that the majority of energy metabolism in this tissue under physiological conditions is anaerobic, assuming that there is no enhanced glucose or glycogen utilization (Pasteur effect) and associated heat production under imposed anoxia, an assumption justified by the data in Ellington & Hammen (1977). This situation is in accord with the established low aerobic capacity of holothuroid longitudinal muscle (Mattisson 1959, Ellington & Hammen 1977, Hill *et al.* 1978). Ellington & Hammen reported that *Sclerodactyla briareus* muscle used 0.851 μ mol glucose · g wet weight^{-1} · h^{-1} under anoxia, and that 94% of the recovered radioactivity was in alanine and lactate. According to the relationships in Gnaiger (1980) for the heat equivalents of dissipative ATP turnover associated with these endproducts, these data indicate a $\dot{Q}_H$ of between 0.057 and 0.087 J · g^{-1} · h^{-1}. Anoxic $\dot{Q}_H$ measured directly in longitudinal muscle of *H.forskali* at the same temperature was 0.079 J · g^{-1} · h^{-1} (Table 4).
muscle of *H.forskali* at the same temperature was 0.079 J · g^{-1} · h^{-1} (Table 4).

Collectively, the direct and indirect calorimetric determinations confirm that the low rate of aerobic energy production by echinoderm tissues under conditions simulating those *in vivo* is supplemented by anaerobic metabolism. By comparison, $\dot{Q}_H$ and $\dot{V}_{O_2}$ in an intertidal sea anemone (*Actinia equina*) and bivalve (*Cardium edule*) are about four times greater than those in echinoderm tissues, and moreover, indicate that these animals remain fully aerobic even during emersion into air (Shick 1981, Shick & Widdows 1981). The failure of exposure of *Patiria pectinifera* and *Strongylocentrotus nudus* to progressive hypoxia ($C_{O_2} < 1$ ml/l) to produce significant changes in the lactate and ATP contents of ovary, pyloric caecum, and gut tissues (Ryabushko *et al.* 1980) suggests that much of the metabolism in these tissues is anaerobic even under environmental normoxia.

Again, a substantial anaerobic contribution to total energy metabolism in the massive internal tissues of echinoderms represents a quantitatively important, but difficult to measure, term in energy budgets. In holothuroids such metabolism involves the production and release of lactate, which may be involved in the generally more acidic nature of holothuroid perivisceral fluid, and the 'dumping' of lactate may also contribute further to the discrepancies in the energy budgets of other echinoderms.

Finally, it is of some interest that chronically hypoxic echinoderm tissues utilize the lactate pathway rather than the succinate or 'opine' pathways, which are, respectively, more efficient energetically and in maintaining NAD+/NADH ratios (De Zwaan 1977, Zandee *et al.* 1980, Fields & Quinn 1981). It may be that once the phylogenetically ancient lactate pathway was inherited from their very early ancestors (Livingstone, 1983), selection in the echinoderms has acted to modify it according to their particular situation (*e.g.* low activity and very high K_m of LDH of internal tissues in the direction of lactate oxidation; reduction of energy demand by tissues and lack of a Pasteur effect under anoxia). Additionally, the presence of other anaerobic pathways, at least in some tissues, has not definitively been ruled out (Ellington 1982).

7 SUMMARY AND PROSPECTS

A large background of ecological, morphological, and biochemical information has identified target species for functional analysis of respiratory gas exchange at the systemic and organismal levels of organization. Adoption of the infaunal lifestyle by some forms has required adaptations to generate respiratory currents (rare among the echinoderms), enabling the study of the role of convective water flow in gas exchange in these species. Basic differences in body plans are associated with different convection mechanisms, and a comparative investigation of them has proved interesting from the standpoints of biomechanics, physiology, and ecology. Anatomical and morphological adaptations to promote oxygen uptake in the highly specialized irregular echinoids have only begun to be investigated physiologically. Comparative studies of the morphometry of the water lung in the ecologically and morphologically diverse holothuroids could also prove interesting. The occurrence of hemoglobin in the holothuroids has been a boon to comparative biochemists, but its exact physiological role in the organisms remains to be shown convincingly; needed are further studies of red cell physiology and biochemistry, and determinations of P_{O_2} at the sites of oxygen unloading. The simultaneous operation of aerobic and anaerobic metabolism in internal tissues poses problems not only in measuring energy turnover in echinoderms, but also in assessing the determinants of pH in their body fluids, which contain acidic endproducts of anaerobiosis. Direct metabolic calorimetry promises to be a powerful technique in determining the aerobic/anaerobic balance, but the size of measuring chambers in present instruments limits its application to small individuals or isolated tissues. The acid-base status of echinoderm body fluids has scarcely been investigated since the advent of modern analytical techniques. Since the mass of internal organs shows large temporal changes in many echinoderms, a coordinated seasonal study of organismal gas exchange, tissue metabolism, and acid-base balance could be very enlightening.

8 ACKNOWLEDGEMENTS

Many colleagues have graciously provided their own unpublished data, references of which I was unaware, or helpful discussions: L.E.Burnett, P.J.Butler, A.M.Clark, J.H.Dearborn, J.A.Dykens, W.R.Ellington, R.H.Emson, W.B.Gladfelter, D.F.Houlihan, K.Johansen, J.M.Lawrence, D.R.Livingstone, J.M.Nance, D.Nichols, D.L.Pawson, and especially to the late J.A.Petersen. Much of the writing and some original research reported here were done while I was on sabbatical at the Institute for Marine Environmental Research, Plymouth, England, where B.L.Bayne and J.Widdows provided laboratory facilities and stimulating discussion, and D.Bale gave library and translation assistance. M.Whitfield and A.D. Mattacola of the Marine Biological Association, Plymouth, kindly supplied a computer program for the acid-base status of seawater, and live echinoderms, respectively. The library of the MBA was an invaluable source of references. I also thank the Harttop family, J.Jennings and associates, and R.Oak for their considerable input during my stay in Plymouth. J.M.Lawrence and C.P.Mangum commented on the entire manuscript, for which I am very grateful. I especially thank J.M.Lawrence for suggesting the topic of this review. Ms E.Albright typed the manuscript. My own research on echinoderms has been supported by US National Science Foundation grants PCM76-81467 and PCM79-11027 (Regulatory Biology), and SPI80-13119 (Science Faculty Professional Development).

9 REFERENCES

Alexander, D.E. & J.Ghiold 1980. The functional significance of the lunules in the sand dollar, *Mellita quinquiesperforata*. *Biol. Bull.* 159:561-570.

Bell, B.M. 1977. Respiratory schemes in the class Edrioasteroidea. *J. Paleontol.* 51:619-632.

Belman, B.W. & A.C.Giese 1974. Oxygen consumption of an asteroid and an echinoid from the Antarctic. *Biol. Bull.* 146:157-164.

Bianconcini, M.S.C., L.O.Medeiros, L.F.Medeiros, E.G.Mendes & D.Valente 1980. Glycolytic and hexosemonophosphate enzyme activities in the lantern muscles of the sea urchins, *Arbacia lixula* (Linn.), *Echinometra lucunter* (Linn.) and *Lytechinus variegatus* (Lamarck). *Comp. Biochem. Physiol.* 67B:569-573.

Binyon, J. 1962. Ionic regulation and mode of adjustment to reduced salinity of the starfish *Asterias rubens* L. *J. mar. biol. Ass. U.K.* 42:49-64.

Binyon, J. 1972. *Physiology of echinoderms.* Pergamon Press, New York. 264pp.

Brown, W.I. 1978. Bimodal gas exchange and the regulation of O_2 uptake in holothurians. M.Sc. thesis, University of Maine, Orono. 56pp.

Brown, W.I. & J.M.Shick 1979. Bimodal gas exchange and the regulation of oxygen uptake in holothurians. *Biol. Bull.* 156:272-288.

Cameron, J.N. 1979. Excretion of CO_2 in water-breathing animals – a short review. *Mar. Biol. Letts* 1:3-13.

Chesher, R.H. 1963. The morphology and function of the frontal ambulacrum of *Moira atropos* (Echinoidea: Spatangoida). *Bull. mar. Sci.* 13:549-573.

Chesher, R.H. 1969. Contributions to the biology of *Meoma ventricosa* (Echinoidea: Spatangoida). *Bull. mar. Sci.* 19:72-110.

Christensen, A.M. 1970. Feeding biology of the sea-star *Astropecten irregularis* Pennant. *Ophelia* 8:1-134.

Clark, R.B. 1964. *Dynamics in metazoan Evolution.* Oxford Univ. Press, London. 313 pp.

Colacino, J.M. 1973. A mathematical model of oxygen transport in the tube foot-ampullar system of the sea-cucumber, *Thyone briareus* L. Ph.D. Dissert., State Univ. New York, Buffalo. 143pp.

Cole, R.N. & W.W.Burggren 1981. The contribution of respiratory papulae and tube feet to oxygen uptake in the sea star, *Asterias forbesi* (Desor.). *Mar. Biol. Letts* 2:279-287.

Cole, W.H. 1940. The composition of fluids and sera of some marine animals and of the seawater in which they live. *J. gen. Physiol.* 23:575-584.

Collip, J.B. 1920. The alkali reserve of marine fish and invertebrates. The excretion of carbon dioxide. *J. biol. Chem.* 44:329-344.

Crozier, W.J. 1916. The rhythmic pulsations of the cloaca of holothurians. *J. exp. Zool.* 20:297-356.

Dejours, P. 1975. *Principles of comparative respiratory Physiology.* North-Holland Publishing Co., Amsterdam. 253pp.

Dejours, P. 1979. Oxygen demand and gas exchange. In: S.C.Wood & C.Lenfant (eds), *Evolution of respiratory Processes: a comparative Approach,* pp.1-49. Marcel Dekker, Inc., New York.

Des Arts, L. 1910. Uber die Lebensweise von *Amphiura chiajei* unter Beruchsichtigung der anatomischen Verhaltnisse. *Bergens Mus. Arb.* 12:1-10.

De Zwaan, A. 1977. Anaerobic energy metabolism in bivalve molluscs. *Oceanogr. mar. Biol. ann. Rev.* 15:103-187.

Diehl, W.J. III, L.McEdward, E.Proffitt, V.Rosenberg & J.M.Lawrence 1979. The response of *Luidia clathrata* (Echinodermata: Asteroidea) to hypoxia. *Comp. Biochem. Physiol.* 62A:669-671.

Dimock, R.V. jr. 1977. Effects of evisceration on oxygen consumption by *Stichopus parvimensis* Clark (Echinodermata: Holothuroidea). *J. exp. mar. Biol. Ecol.* 28:125-132.

Durham, J.W. 1966. Clypeasteroids. In: R.C.Moore (ed.), *Treatise on Invertebrate Paleontology.* Part U, Echinodermata 3, Vol.2, pp.450-491.

Ellington, W.R. 1975. Glucose degradation and respiratory metabolism in starfish tissue. *Fed. Proc.* 34:466.
Ellington, W.R. 1982. Intermediary metabolism. In: M.Jangoux & J.M.Lawrence (eds), *Echinoderm Nutrition,* pp.395-415. Balkema, Rotterdam.
Ellington, W.R. & C.S.Hammen 1977. Metabolic compensation to reduced oxygen tensions in the sea cucumber, *Sclerodactyla briareus. J. comp. Physiol.* 122:347-358.
Engstrom, N.A. 1982. Brooding behavior and reproductive biology of a subtidal Puget Sound sea cucumber, *Cucumaria lubrica* (Clark 1901) (Echinodermata: Holothuroidea). In: J.M.Lawrence (ed.), *Echinoderms: Proc. Intern. Conf., Tampa Bay.* pp.447-450. Balkema, Rotterdam
Farmanfarmaian, A. 1966. The respiratory physiology of echinoderms. In: R.A.Boolootian (ed.), *Physiology of Echinodermata,* pp.245-265. John Wiley & Sons, New York.
Fenner, D.H. 1973. The respiratory adaptations of the podia and ampullae of echinoids (Echinodermata). *Biol. Bull.* 145:323-339.
Ferber, I. 1976. Functional morphology of *Lovenia elongata* (Gray) (Echinoidea: Spatangoida). *Thalass. Jugoslav.* 12:123-128.
Ferber, I. & J.M.Lawrence 1976. Distribution, substratum preference, and burrowing behavior of *Lovenia elongata* (Gray) (Echinoidea: Spatangoida) in the Gulf of Elat (Aqaba), Red Sea. *J. exp. mar. Biol. Ecol.* 22:207-225.
Field, J.G. 1972. Some observations on the release of dissolved organic carbon by the sea urchin *Strongylocentrotus droebachiensis. Limnol. Oceanogr.* 17:759-761.
Fields, J.H.A. & J.F.Quinn 1981. Some theoretical considerations on cytosolic redox balance during anaerobiosis in marine invertebrates. *J. theor. Biol.* 88:35-45.
Foster-Smith, R.L. 1978. An analysis of water flow in tube-living animals. *J. exp. mar. Biol. Ecol.* 34:73-95.
Gentleman, S.B. 1970. Respiratory Studies of a Sea Cucumber. Ph.D. Dissert., Duke Univ. 53pp.
Giese, A.C. 1966. On the biochemical constitution of some echinoderms. In: R.A.Boolootian (ed.), *Physiology of Echinodermata,* pp.757-796. John Wiley & Sons, New York.
Giese, A.C. 1976. Physiology of the echinoderm body wall. *Thalass. Jugoslav.* 12:153-163.
Giese, A.C., A.Farmanfarmaian, S.Hilden & P.Doezema 1966. Respiration during the reproductive cycle in *Strongylocentrotus purpuratus. Biol. Bull.* 130:192-201.
Gislén, T. 1924. Echinoderm studies. *Zool. Bidr. Uppsala* 9:1-316.
Gladfelter, W.B. 1978. General ecology of the cassiduloid urchin *Cassidulus cariboearum. Mar. Biol.* 47:149-160.
Gnaiger, E. 1980. Das kalorische Aquivalent des ATP- Umsatzes im aeroben und anoxischen Metabolismus. *Thermochim. Acta* 40:195-223.
Greenwood, P.J. 1980. Growth, respiration and tentative energy budgets for two populations of the sea urchin *Parechinus angulosus* (Leske). *Estuar. coast. mar. Sci.* 10:347-367.
Hansen, B. 1975. Systematics and biology of the deep-sea holothurians. Part I. Elasipoda. *Galathea Rep.* 13:1-262.
Haugh, B.N. 1973. Water vascular system of the Crinoidea Camerata. *J. Paleontol.* 47:77-90.
Herreid, C.F.II, V.F.LaRussa & C.R.DeFesi 1976. Blood vascular system of the sea cucumber, *Stichopus moebii. J. Morph.* 150:423-452.
Herrera, F.C. & M.Plaza 1981. Extracellular cations and respiration of tissues of *Holothuria glaberrima. Comp. Biochem. Physiol.* 70A:27-32.
Hiestand, W.A. 1940. Oxygen consumption of *Thyone briareus* (Holothuroidea) as a function of oxygen tension and hydrogen-ion concentration of the surrounding medium. *Trans. Wisc. Acad. Arts Lett.,* 32:167-175.

Higgins, R.C. 1974. Observations on the biology of *Apatopygus recens* (Echinoidea: Cassiduloida) around New Zealand. *J. Zool., Lond.* 173:505-516.
Hill, R.B., J.W.Sanger, R.E.Yantorno & C.Deutsch 1978. Contraction in a muscle with negligible sarcoplasmic reticulum: the longitudinal retractor of the sea cucumber *Isostichopus badionotus* (Selenka), Holothuria Aspidochirota. *J. exp. Zool.* 206:137-150.
Houlihan, D.F. & G.Duthie 1981. Measurement of oxygen consumption and sampling of body fluids of echinoderms in situ. *J. exp. mar. Biol. Ecol.* 51:97-106.
Howell, B.J., H.Rahn, D.Goodfellow & C.Herreid 1973. Acid-base regulation and temperature in selected invertebrates as a function of temperature. *Amer. Zool.* 13:557-563.
Hyman, L.H. 1955. *The Invertebrates, vol.IV: Echinodermata.* McGraw-Hill Inc., New York. 763pp.
Jöbsis, F.F. (ed.), 1977. *Oxygen and Physiological Function.* Professional Information Library, Dallas, Texas. 550pp.
Johansen, K. & J.A.Petersen 1971. Gas exchange and active ventilation in a starfish, *Pteraster tesselatus. Z. vergl. Physiol.* 71:365-381.
Johansen, K. & R.L.Vadas 1967. Oxygen uptake and responses to respiratory stress in sea urchins. *Biol. Bull.* 132:16-22.
Kawamoto, N. 1928. Oxygen capacity of the blood of certain invertebrates which contain haemoglobin. *Sci. Rep. Tohôku Univ.* (4)3:561-575.
Kier, P.M. 1962. Revision of the cassiduloid echinoids. *Smithsonian misc. Coll.* 144:1-262.
Kier, P.M. 1974. Evolutionary trends and their functional significance in the post-Paleozoic echinoids. *J. Paleontol.* 48(Suppl. to No.3: Paleontol. Soc. Mem. 5):1-95.
Lawrence, J.M. & J.E.M.Lane 1982. The utilization of nutrients by post-metamorphic echinoderms. In: M.Jangoux & J.M.Lawrence (eds), *Echinoderm Nutrition,* pp.331-371. Balkema, Rotterdam.
Lewis, J.B. 1968. Comparative respiration of tropical echinoids. *Comp. Biochem. Physiol.* 24:649-652.
Lilly, G.R. 1979. The influence of diet on the oxygen uptake of the sea urchins, *Tripneustes ventricosus* and *Strongylocentrotus droebachiensis. Comp. Biochem. Physiol.* 62A: 463-470.
Livingstone, D.R. 1983. Invertebrate and vertebrate pathways of anaerobic metabolism: evolutionary considerations. *J.geol. Soc.* 140:27-38.
Madsen, F.J. 1961. The Porcellanasteridae. A monographic revision of an abyssal group of sea stars. *Galathea Rep.* 4:33-174.
Maloeuf, N.S.R. 1937. Studies on the respiration (and osmoregulation) of animals. I. Aquatic animals without an oxygen transporter in their internal medium. *Z. vergl. Physiol.* 25:1-28.
Mangum, C.P. 1973. Evaluation of the functional properties of invertebrate hemoglobins. *Neth. J. Sea Res.* 7:303-315.
Mangum, C.P. 1976. Primitive respiratory adaptations. In: R.C.Newell (ed.), *Adaptation to Environment: Essays on the Physiology of Marine Animals,* pp:191-278. Butterworth's, London.
Mangum, C.P. & J.M.Shick 1972. The pH of body fluids of marine invertebrates. *Comp. Biochem. Physiol.* 42A:693-697.
Mangum, C.P. & W.Van Winkle 1973. Responses of aquatic invertebrates to declining oxygen conditions. *Amer. Zool.* 13:529-541.
Manwell, C. 1959. Oxygen equilibrium of *Cucumaria miniata* hemoglobin and the absence of the Bohr effect. *J. cell. comp. Physiol.* 53:75-83.
Manwell, C. 1964. Chemistry, genetics and function of invertebrate respiratory pigments – configurational changes and allosteric effects. In: F.Dickens & E.Neil (eds), *Oxygen in the animal Organism,* pp.49-119. Macmillan, New York.

Manwell, C. 1966. Sea cucumber sibling species: polypeptide chain types and oxygen equilibrium of hemoglobin. *Science* 152:1393-1396.

Mattisson, A.G.M. 1959. Cytochrome c, cytochrome oxidase, and respiratory intensity in some types of invertebrate muscles. *Ark. Zool.* 12:143-163.

Miller, R.J. & K.H.Mann 1973. Ecological energetics of the seaweed zone in a marine bay on the Atlantic coast of Canada. III. Energy transformations by sea urchins. *Mar. Biol.* 18:99-114.

Montuori, A. 1913. Les processus oxydatifs chez les animaux marins en rapport avec la loi de superficie. *Arch. Ital. Biol.* 59:213-234.

Moore, H.B. & B.F.McPherson 1965. A contribution to the study of the productivity of the urchins *Tripneustes esculentus* and *Lytechinus variegatus*. *Bull. mar. Sci.* 15:855-871.

Nance, J.M. & L.F.Braithwaite 1981. Respiratory water flow and production of mucus in the cushion star, *Pteraster tesselatus* Ives (Echinodermata: Asteroidea). *J. exp. mar. Biol. Ecol.* 50:21-31.

Newell, R.C. & W.A.M.Courtney 1965. Respiratory movements in *Holothuria forskali* Delle Chiaje. *J. exp. Biol.* 42:45-57.

Nichols, D. 1959a. Changes in the Chalk heart-urchin *Micraster* interpreted in relation to living forms. *Phil. Trans. roy. Soc.* (B)242:347-437.

Nichols, D. 1959b. The histology of the tube-feet and clavulae of *Echinocardium cordatum*. *Quart. J. micr. Sci.* 100:73-87.

Nichols, D. 1959c. The histology and activities of the tube-feet of *Echinocyamus pusillus*. *Quart. J. micr. Sci.* 100:539-555.

Nomura, S. 1926. The influence of oxygen tension on the rate of oxygen consumption in *Caudina*. *Sci. Rep. Tohôku Univ.* (4)2:133-138.

Ogawa, S. 1928. A new apparatus used for the study of respiration in *Caudina chilensis*. *Sci. Rep. Tohôku Univ.* (4)3:39-49.

Okazaki, K. & T.Koizumi 1926. Uber die Liebeschohlenflussigkeit von Holothurien, *Caudina chilensis* (J.Muller). *Sci. Rep. Tohôku Univ.* (4)2:139-142.

Ott, J. & K.Fedra 1977. Stabilizing properties of a high-biomass benthic community in a fluctuating ecosystem. *Helgolander wiss. Meeresunters.* 30:485-494.

Pamatmat, M.M. 1979. Anerobic heat production of bivalves *(Polymesoda caroliniana* and *Modiolus demissus)* in relation to temperature, body size and duration of anoxia. *Mar. Biol.* 53:223-229.

Pantin, C.F.A. & P.Sawaya 1953. Muscular action in *Holothuria grisea*. *Bol. Fac. Filos. Cienc. Univ. S.Paulo* (Zool.)18:51-59.

Paul, C.R.C. 1968. Morphology and function of dichoporite pore-structures in cystoids. *Palaeontology* 11:697-730.

Paul, C.R.C. 1972. Morphology and function of exothecal pore-structures in cystoids. *Palaeontology* 15:1-28.

Paul, C.R.C. 1976. Respiration rates in primitive (fossil) echinoderms. *Thalass. Jugoslav.* 12:277-286.

Paul, C.R.C. 1977. Evolution of primitive echinoderms. In: A.Hallam (ed.), *Patterns of Evolution as Illustrated by the Fossil Record*, pp.123-158. Elsevier, Amsterdam.

Pearse, A.S. 1908. Observations on the behavior of the holothurian, *Thyone briareus* (Lesueur). *Biol. Bull.* 15:259-288.

Pentreath, R.J. 1971. Respiratory surfaces and respiration in three New Zealand intertidal ophiuroids. *J. Zool., Lond.* 163:397-412.

Percy, J.A. 1974. Thermal adaptation in the boreo-arctic echinoid, *Strongylocentrotus droebachiensis* (O.F.Muller 1776). III. Seasonal acclimatization and metabolism of tissues in vitro. *Physiol. Zool.* 47:59-67.

Petersen, J.A. 1976. Aspects of gas exchange in ophiuroids from the coast of Brazil. *Thalass. Jugoslav.* 12:295-296.
Phelan, T.F. 1977. Comments on the water vascular system, food grooves, and ancestry of the clypeasteroid echinoids. *Bull. mar. Sci.* 27:400-422.
Punzo, F. 1977. The pH of body fluids from marine intertidal invertebrates. *J. exp. mar. Biol. Ecol.* 30:327-331.
Robertson, D.A. 1972. Volume changes and oxygen extraction efficiency in the holothurian, *Stichopus mollis* (Hutton). *Comp. Biochem. Physiol.* 43A:795-800.
Russo, A.R. 1977. Water flow and the distribution and abundance of echinoids (genus *Echinometra*) on an Hawaiian reef. *Aust. J. mar. freshwater Res.* 28:693-702.
Ryabushko, V.I., A.A.Zhuchikhina & N.V.Lutsik 1980. Effects of environmental oxygen concentrations on the level of metabolism of some echinoderms from the Sea of Japan. *Comp. Biochem. Physiol.* 67B:171-174.
Sabourin, T.D. & W.B.Stickle 1981. Effects of salinity on respiration and nitrogen excretion in two species of echinoderms. *Mar. Biol.* 65:91-99.
Sarch, M.N. 1931. Die Pufferung der Korperflussigkeiten bei Echinodermen. *Z. Physiol.* 14:525-545.
Shick, J.M. 1976. Physiological and behavioral responses to hypoxia and hydrogen sulfide in the infaunal asteroid *Ctenodiscus crispatus. Mar. Biol.* 37:279-289.
Shick, J.M. 1981. Heat production and oxygen uptake in intertidal sea anemones from different shore heights during exposure to air. *Mar. Biol. Lett.* 2:225-236.
Shick, J.M., K.C.Edwards & J.H.Dearborn 1981. Physiological ecology of the deposit-feeding sea star *Ctenodiscus crispatus:* ciliated surfaces and animal-sediment interactions. *Mar. Ecol. Prog. Ser.* 5:165-184.
Shick, J.M. & J.Widdows 1981. Direct and indirect calorimetric measurement of metabolic rate in bivalves during aerial exposure. *Amer. Zool.* 21:983
Shumway, S.E. 1977. The effects of fluctuating salinities on four species of asteroid echinoderms. *Comp. Biochem. Physiol.* 58A:177-179.
Skirrow, G. 1975. The dissolved gases – carbon dioxide. In: J.P.Riley & G.Skirrow (eds), *Chemical oceanography,* Vol.2 (2nd ed.), pp.1-192. Academic Press, London.
Sladen, W.P. 1883. The Asteroidea of H.M.S. 'Challenger' Expedition – Part II. *J. Linn. Soc., Zool.* 17:214-269.
Smith, A.B. 1978. A functional classification of the coronal pores of regular echinoids. *Palaeontology* 21:759-789.
Smith, A.B. 1980. The structure, function, and evolution of tube feet and ambulacral pores in irregular echinoids. *Palaeontology* 23:39-83.
Steen, J.B. 1965. Comparative aspects of the respiratory gas exchange of sea urchins. *Acta physiol. Scand.* 63:164-170.
Stewart, P.A. 1978. Independent and dependent variables of acid-base control. *Resp. Physiol.* 33:9-26.
Terwilliger, R.C. & K.R.H.Read 1972. The hemoglobin of the holothurian echinoderm, *Molpadia oolitica* Pourtales. *Comp. Biochem. Physiol.* 42B:65-72.
Theede, H., A.Ponat, K.Hiroki & C.Schlieper 1969. Studies on the resistance of marine bottom invertebrates to oxygen-deficiency and hydrogen sulphide. *Mar. Biol.* 2:325-337.
Thum, A.B. & J.C.Allen 1976. Reproductive ecology of the lamp urchin (*Echinolampas crassa* (Bell, 1880) from a subtidal biogenous ripple train. *Trans. roy. Soc. S.Afr.* 42: 23-33.
Timko, P.L. 1976. Sand dollars as suspension feeders: a new description of feeding in *Dendraster excentricus. Biol. Bull.*, 151:247-259.
Ultsch, G.R. & G.Gros 1979. Mucus as a diffusion barrier to oxygen: possible role in O_2

uptake at low pH in carp *(Cyprinus carpio)* gills. *Comp. Biochem. Physiol.* 62A:685-689.

Webster, S.K. 1975. Oxygen consumption in echinoderms from several geographical locations, with particular reference to the Echinoidea. *Biol. Bull.* 148:157-164.

Webster, S.K. & A.C.Giese 1975. Oxygen consumption of the purple sea urchin with special reference to the reproductive cycle. *Biol. Bull.* 148:165-180.

Winterstein, H. 1909. Uber die Atmung der Holothurien. *Arch. Fisiol.* 7:87-93.

Wolcott, T.G. 1981. Inhaling without ribs: the problem of suction in soft-bodied invertebrates. *Biol. Bull.* 160:189-197.

Woodley, J.D. 1975. The behaviour of some amphiurid brittle-stars. *J. exp. mar. Biol. Ecol.* 18:29-46.

Zammit, V.A. & E.A.Newsholme 1976. The maximum activities of hexokinase, phosphorylase, phosphofructokinase, glycerol phosphate dehydrogenases, lactate dehydrogenase, octopine dehydrogenase, phosphoenolpyruvate carboxykinase, nucleoside diphosphatekinase, glutamate-oxaloacetate transaminase and arginine kinase in relation to carbohydrate utilisation in muscles from marine invertebrates. *Biochem. J.* 160:447-462.

INNATE AND LEARNED RESPONSES TO EXTERNAL STIMULI IN ASTEROIDS

TINE VALENTINČIČ
Institute & Department of Biology, University of Ljubljana, Yugoslavia

CONTENTS

1 INTRODUCTION

There have been only seven papers presenting new observations on learning in echinoderms since 1940 (Sokolov 1961, Landenberger 1966, 1968, Castilla 1972a, Ormond *et al.* 1976, Valentinčič 1978, 1980a). At least five reviews on the same topic appeared during the same period (Hyman 1955, Thorpe 1956, Reese 1966, Binyon 1972, Willows & Corning 1975) and additional information is included in the review by Sloan & Campbell (1982) on perception of food. Applewhite (1979) wrote in his review of learning in Protozoa: 'I like protozoa but they do not like me. They have infected me, and they behave badly.' A similar conclusion may be applicable working with echinoderm behavior and learning. Radial symmetry, relatively slow locomotion, unusual locomotory device (tube feet), a unique nervous system, and lack of knowledge of sensory capacities make experimental manipulation and learning studies extremely difficult. Not a single technique developed for learning studies in vertebrates seems to be transferable to echinoderms.

Thorpe (1956) and Reese (1966) concluded that habituation had been demonstrated in echinoderms. Their statements were based mainly on Preyer's (1887) and Ven's (1921) works. In the most careful studies of Ven, individual *Asterias rubens* were fixed with pegs or collars across the arms and the escape time to a 'familiar environment' was measured. The persistence of ineffective movements and the escape time decreased. These changes cannot be explained as a simple habituation – a waning of the response which has no biological meaning to the animals. We still lack an elegant experiment demonstrating habituation in echinoderms in the same way that Jennings described it in the sea anemone *Aiptasia annulata* under the label, acclimatization to stimuli, as long ago as 1906.

The work of Jennings (1907) differed from that of Preyer and Ven. Jennings was able to modify the behavior of an asteroid in the righting reflex. The least preferred arms frequently were used to initiate the righting response after punishment of the dominant arms. Most authors wishing to repeat these experiments found the experimental situation extremely complicated and the results were often contradictory or 'negative'. The inventory of possible asteroid movements was large and the conditions of stimulation changed from trial to trial.

2 RESPONSIVENESS OF QUIESCENT AND ACTIVE ASTEROIDS TO EXTERNAL STIMULI

Asteroids are inactive during most of the 24 hour daily cycle. Periods of daily rhythmic activities occur in different species of *Astropecten* (Mori & Mutatani 1952, Massé 1975). The activity rhythm was lost in constant darkness and Fenchell (1965) believed that the circadian behavior is controlled by a photoperiod. To my knowledge there are no published studies demonstrating an endogenous nature of circadian rhythms in asteroids under laboratory conditions.

The activity of inactive individuals may be triggered by chemical, mechanical, and light stimuli. Very strong stimuli might provoke escape behavior. Minor stimuli signal the presence of food, predators, other individuals, and sex partners. Motivational states often influence the responsiveness of animals to key stimuli. Romanes (1983) and Preyer (1887) were the first to notice that only hungry asteroids responded to feeding stimuli. By increasing the level of motivation, behavioral thresholds for key stimuli are lowered and responses to unspecific stimuli may occur. Vacuum activity (Lorenz 1935), also termed overflow activity (Bastock *et al.* 1953), occurs in the asteroid *Marthasterias glacialis* after extreme deprivation of food (Valentinčič, unpublished).

In the asteroids with well-developed suckers on their tube feet, at least two states of inactivity are virtually identical in appearance. Individuals may be loosely attached to the substrate and can be easily detached without damaging the tube feet. Individuals also may be firmly attached to the substrate and cannot be detached without breaking the tube feet. Inactivity thus includes at least two different states with all possible intermediates.

With only few exceptions (Zafiriou *et al.* 1972, Valentinčič 1973, 1975, 1980b,c) most authors have studied responsiveness of active individuals to external stimuli (Reese 1966, Feder & Christensen 1966, Yoshida 1966, Millot 1966, Binyon 1972, Sloan 1980, Sloan & Campbell 1982). Undisturbed animals were observed only on rare occasions (Barnes *et al.* 1970, Brauer & Jordan 1970). Unfortunately, experimental individuals were often manipulated at the beginning of the experiments. Flow tanks and Y-mazes were employed for reasons of a 'binomial simplicity'. The most extensive studies of active individuals towards external stimuli are those of Castilla (1971, 1972a,b), Castilla & Crisp (1970, 1973), Whittle & Blumer (1970a,b), Zafiriou *et al.* (1972) and Zafiriou (1972). A Y-maze was also used in nature to examine chemoreception of *Acanthaster planci* by Ormond *et al.* (1973) and by Ormond & Campbell (1974). A device similar to a Y-maze was used by Sokolov (1961) in tactile conditioning studies in *Asterias rubens.*

The term, active animals, is even more inaccurate than the term, inactive animals. The common feature of many active states is locomotion. Several papers report on temporary dominance of single rays – leading arms during locomotion. Kerkut (1954) concluded that both the central nervous system and external stimuli contribute to the coordinated movements of the tube feet. The tube feet of advancing asteroids are coordinated in such a way to move the individual with the leading arms forward. Jennings (1907) assumed the central origin of the persistence of movement in a certain direction. In the most extreme case of *Pycnopodia helianthoides* (Kjerskog-Agersborg 1918, 1922), the rays close to the madreporite always lead locomotion. Such constant bilaterality is not known in many other asteroids. A kind of temporary functional bilaterality during the movements always exists.

In the learning studies of Sokolov (1960) and in many other works, the experimental manipulation at the beginning of an experiment influence the state of the individual and might also have influenced the direction of ensuing locomotion. *Marthasterias glacialis* can use any of its five arms in leading the locomotion and pairing of any of the adjacent rays is also possible (Preyer 1887, Tortonese 1950). The leading arms during locomotion have their distal tube feet extended and the tips of the arms slightly curled.

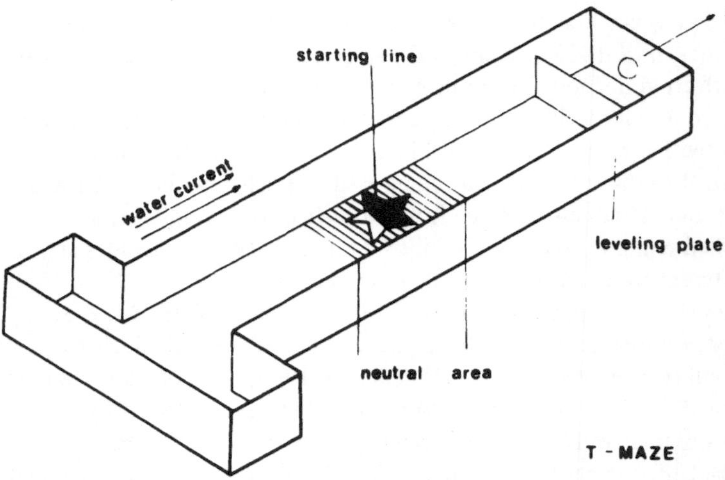

Figure 1. T-maze. The long arm of the maze is used as a flow tank.

3 THE EXPERIMENTAL MANIPULATION OF DOMINANT AND SUBORDINATE ARMS AND THE BEHAVIOR OF ASTEROIDS IN FLOW TANKS

A series of experiments were done in our laboratory to discover the consequences of the experimental manipulation on the behavior of *Marthasterias glacialis.* The experiments which have potential for bioassay techniques and learning studies will be presented here. *Marthasterias glacialis* collected in the Gulf of Rijeka were maintained in a condition of hunger to show appetitive feeding responses to stimulation with 10^{-6} M L-cysteine. The time-consuming technique of controlled starvation (Valentinčič 1973) was avoided. A suitable feeding motivation level was obtained by collecting individuals during winter or spring periods. An increase in the aquaria temperature from 10-15°C to 22°C affected the hunger level. Individuals started to respond to L-cysteine stimulation within 10 days. Those individuals were fed later at the level to maintain the high responsiveness to the chemical stimulation.

The long arm of a T-maze was used as a flow tank (Fig.1). The sea-water velocity in the flow tank was 14 mm/s, well above the threshold for the rheotactic response (Castilla & Crisp 1973). Individuals were transferred from the storage tanks into the intermediate aquarium and used for experiments during periods of locomotion. Four types of experiments were carried out on the same group:

– Subordinate-arms manipulated group (SA): one of the subordinate arms

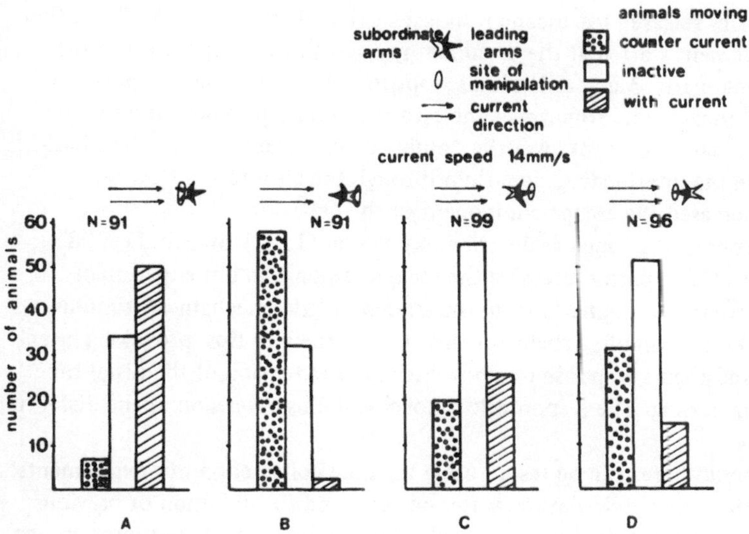

Figure 2. The direction of movement of *Marthasterias glacialis* in the flow tank after manipulation of a dominant arm (LA) and after manipulation of a subordinate arm (SA) (Valentinčič & Dolenc, unpublished).

was grasped and the individual transferred onto the start line of the flow channel either with the dominant (leading) arms in the direction of the current – SA/WC, or with the dominant arms in the direction opposite the current (counter-current) – SA/CC.

– Dominant-arms manipulated group (LA): one of the dominant arms was grasped and the individual placed onto the start line either with the dominant arms in the direction of the current – LA/WC, or with the dominant arms in the direction opposite the current (counter-current) – LA/CC.

The results agreed with Jenning's view of the persistence of movement. The SA individuals moved in the direction of dominant arms independently of the direction of water current (Fig.2A,B). Many LA individuals became inactive (Fig.2C,D). A considerable number changed the direction of movement to the direction of the arms opposite the manipulated one.

The experimental manipulation of an arm induced tube-feet retraction, arm shortening, elevation of the affected ray, and escape in the direction opposite to the touch. This response is more or less identical with the escape response as described by Valentinčič (1978, 1980). The manipulation of the subordinate arms reinforced the previously established direction of movement. The escape response was initiated at the affected subordinate ray, and the individuals escaped in the direction of the dominant arms. The manipulation of the dominant arms released the escape response at the affected arm tip. The redirection of movement started at the dominant ray and slowly advanced towards the subordinate arm. Many individuals stopped before resuming the previous direction of movement. Sometimes the escape response prevailed and the previous subordinate arms took over the dominant role. The removal of the individuals from the intermediate aquarium through the air into the flow tank probably increased the escape component of the behavior.

The responses to a touch as described by Prayer (1887), Mangold (1908) and Kalmus (1929) partly resemble the escape response in our experiments. The main difference originates from the strength of the mechanical stimulus. Weak mechanical stimuli, a slight touch at one of the arm tips, provoke a brief retraction and then a response to a novel stimulus-extension of the distal tube feet and arm curling. The response to a novel stimulus extension of the distal tube ing response.

Several hypotheses can be tested using the data collected in our experiments:

– Experimental manipulation, water current, and the direction of previous movements do not affect the results. Half of the individuals should move in one direction and the other half in the opposite direction. The highly significant departure from this expectation in SA individuals ($P \leqslant 0{,}001$) excludes this possibility (Table 1).

– The LA and the SA individuals can be turned in the same direction. The influence of water currents is thus the same for both groups. If the experimental manipulation and the previously established direction of movement do not influence the behavior, there should be no difference between individuals turned in the same direction with respect to current. The highly significant difference ($P \leqslant 0{,}001$) excludes such a possibility (Table 2).

– The experimental manipulation and the previously established direction of movement decisively influence the ensuing result. Under this hypothesis there should be no difference within the groups with the manipulation of arms in the same state in keeping the leading arms dominant. This is the case in our experiments (Table 3).

Table 1. Movement of *Marthasterias glacialis* in the flow tank after the manipulation of a dominant (LA) or subordinate arm (SA). The dominant arms were oriented in the direction of current (WC) or in the direction opposite the current (counter current, CC). (Valentinčič & Dolenc, unpublished).

Experiment	Number of individuals moving counter the current	Number of individuals inactive	Number of individuals moving with the current	X^2
SA/WC	7	34	50	30.95
SA/CC	57	32	2	49.43
LA/WC	19	55	25	0.80
LA/CC	31	51	14	1.42

Table 2. Comparison of the movement of ***Marthasterias glacialis*** after manipulation of a dominant (LA) or subordinate (SA) arm when individuals are oriented in the direction with current (WC) or in the direction opposite the current (counter current, CC). (Valentinčič & Dolenc, unpublished).

Comparison of			X^2
SA/WC	with	LA/WC	10.84
SA/CC	with	LA/CC	13.02

Table 3. Numbers of ***Marthasterias glacialis*** keeping the dominant arms (LA) and changing the dominant arms after different experimental manipulations (Valentinčič & Dolenc, unpublished).

Experiment	Number keeping the dominant arms	Number inactive	Number changing the dominant arms	X^2
SA/WC	50	34	7	2.08
SA/CC	57	32	2	
LA/WC	25	55	19	0.92
LA/CC	31	51	14	

The inactive animals, which were not included in statistical evaluation, frequently attached strongly to the bottom. This behavior was called the 'fright response'.

4 RHEOTAXIS AND CHEMOTAXIS STUDIES IN FLOW TANKS

Feeding stimuli in water currents could be expected to induce an upstream behavior in the SA/WC individuals. This is not the case with key chemical stimuli for feeding (Fig.3). L-cysteine, the strongest feeding-response releasing stimulus (Valentinčič 1973, 1975, 1980a,b), significantly decreases the 'inactivation' of individuals but did not redirect them. Such a result should be expected since the solution in the flow channel was more or less homogeneous. There was no gradient between the front and the rear arms. The data of Whittle & Blumer (1970a) indicate that the distribution of the test stimuli was not homogeneous in their flow tank. There was a gradient between the lateral and middle parts of their tank and a decreasing concentration of the stimuli from the upper part of the tank to the drain. The ensuing positive chemotactic responses could thus be based on chemical stimuli alone. In contrast, mussel extracts and bacteriological peptone provoked at least some upstream movement in SA/WC animals with currents of 14 mm/s and 3 mm/s (Fig.4). Movement at low velocity did not differ from movement in the control. There is no basis for invoking unknown attractants or a combination of chemical stimuli and water currents to explain this upstream movement.

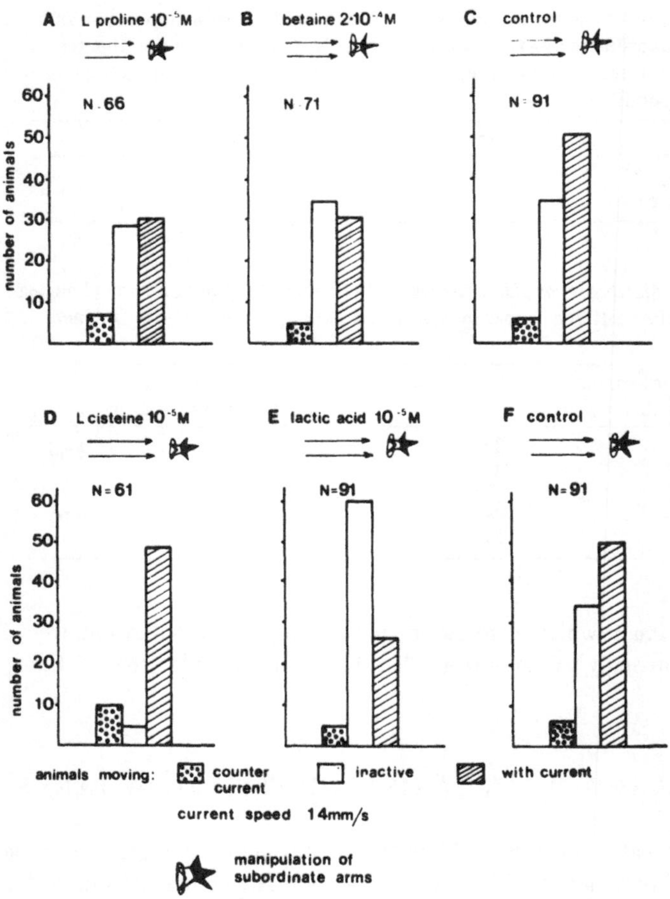

Figure 3. The influence of low molecular weight compounds on downstream movement of *Marthasterias glacialis* (Valentinčič & Dolenc, unpublished).

5 PROBABILITY FOR UPSTREAM MOVEMENT IN *MARTHASTERIAS GLACIALIS* AS A RESULT OF RANDOM EXPERIMENTAL MANIPULATION

'Positive rheotatic' or 'negative rheotactic' behavior can occur as a result of experimental manipulation. The arms manipulated are usually selected at random. Two or three arms may be dominant or subordinate in *Marthasterias glacialis.* For simplicity, one can assume that the probabilities of manipulating dominant or subordinate arms are the same. The data in table 1 and simple probability calculations allow us to predict the direction of movement.

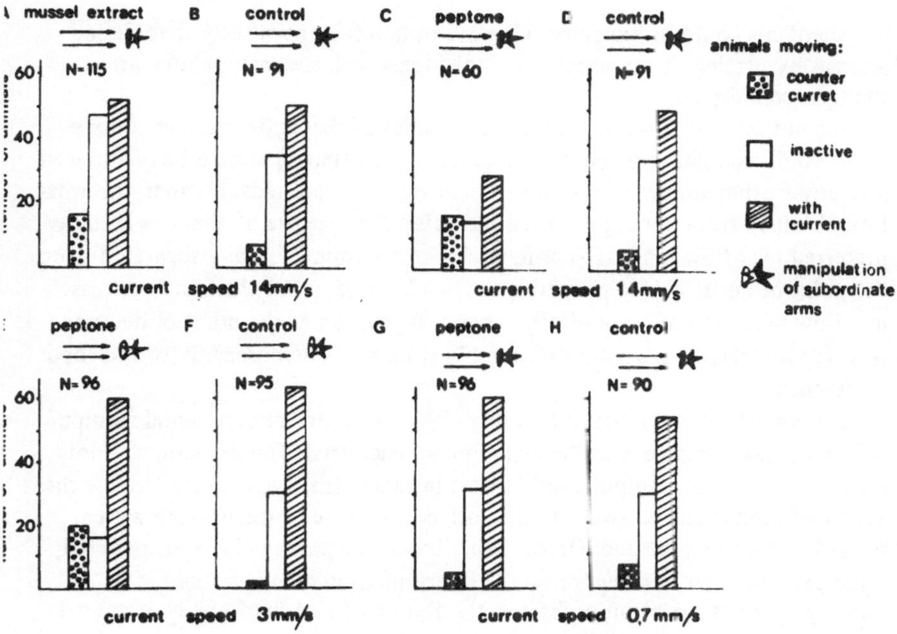

Figure 4. The influence of mussel extract and bacteriological peptone on downstream movement of *Marthasterias glacialis* (Valentinčič & Dolenc, unpublished).

Active individuals placed into the flow tank with the manipulated arm oriented in the downstream direction moved mostly upstream:

$$P(SA) \times P(SA/CC) + P(LA) \times P(LA/WC) = P(U)$$

P(SA): probability of manipulating subordinate arms,

P(SA/CC): probability that the individuals manipulated by subordinate arms and placed into the flow tank with the dominant arms opposite the direction of current (counter-current) will move upstream,

P(LA): probability of manipulating leading arms,

P(LA/WC): probability that the individuals manipulated by dominant arms and placed into the flow tank with the dominant arms downstream will move upstream,

P(U): probability that the individual will move upstream (false positive rheotaxis).

Taking the adequate sets of data, one obtains: $1/2 \times 57/59 + 1/2 \times 19/44 = 0.699$. Seventy per cent of the individuals used in experiments and placed on the start line with the manipulated arms downstream would thus move upstream. All arms tend to be subordinate in inactive animals. As much as 90 % false positive rheotaxis could be expected. One can obtain false negatively rheotactic animals by placing the experimental individuals with the manipulated arm in the upstream direction.

It is not my intention to dismiss past studies of rheotaxis and chemotaxis in asteroids. Necessary precautions regarding the behavior should be introduced into any further studies of responses to currents by asteroids. Further information obtained from our experiments indicated that the use of Y-mazes is to be preferred over flow tanks. Experimental manipulation is usually separated from the point of decision by time and space in a Y-maze, while this is not the case in a flow tank. The Y-maze offers a chemical gradient at the point of decision. It is, for this reason, particularly useful in studies of chemosensitivity and food preferences.

The conditioning studies of Sokolov (1961) included experimental manipulations of *Asterias rubens* at the beginning of each trial. The decision was not separated from the manipulation, at least in space. His study, suggestive for the ability of individuals to associate the surface structure of the bottom with a reward, should be repeated. On the other hand, Diebschlag (1938) conducted his experiments on learning for tactile discrimination of substrates and on learning for discrimination of differently illuminated parts of test containers in a more subtle manner. The experimental manipulation was carried out only at the beginning of each experiment. The successive punishments were given to his experimental individuals after they tried to leave the middle part of the container which was different from the periphery. The punishments were repeated each time the individual tried to leave the permitted area. Thus the successive trials were separated in time from the experimental manipulation. In my opinion, his results, though the number of experiments was small, are the first demonstration of associative learning in echinoderms.

6 PHOTOTACTIC RESPONSES

Data scattered in literature on phototactic responses in asteroids and the Castilla's (1971) report on the changes in *Asterias rubens* from photonegative behavior in the field to photopositive behavior under constant illumination, and indifferent behavior after constant darkness, inspired us to follow the Landenberger's procedure for associative learning (1966). *Pisaster giganteus* were adapted to constant darkness in Landenberg's experiments. Fifteen minutes' illumination and food stimuli were simultaneously presented to the animals during the training trials. Not all the animals responded to the stimulation in the first training trials. The response was 100 % during consecutive training trials and during test trials. The food was offered after the end of the period of illumination in the test trials, but all animals responded slowly to illumination alone. The response waned when no food was presented.

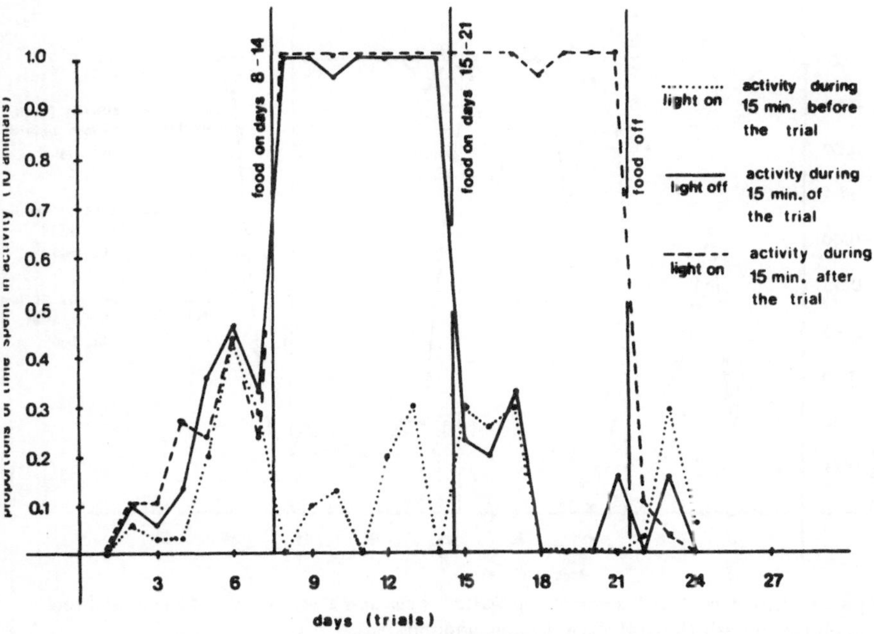

Figure 5. Activities of light-adapted *Marthasterias glacialis* after exposure to darkness and food at 24-hour intervals (Valentinčič & Kranjc, unpublished).

The same behavior as in *Pisaster giganteus* is often observed in *Marthasterias glacialis.* Individuals stationary on the walls descend to the bottom or start to explore the water surface after stimulation with food. One group of ten hungry *Marthasterias glacialis* was adapted to constant 'darkness' (the illumination was below 1 lx) and a second group was adapted to constant illumination (a 100 W ordinary bulb 30 cm above the surface of a 800 1 container). During the first seven days the animals were exposed to 15 minutes of illumination at 24-hour intervals. The activity of individuals was recorded (in terms of minutes) during 15 minutes prior to the onset of the experiments, during the 15 minutes of exposure to light, and for 15 minutes after exposure to light. In days 8 to 14, the light and the food stimuli (mussels) were presented simultaneously. In the light-adapted individuals no difference was found in activities before, during and after the illumination period (Fig.5). A slight increase in activities was detected after illumination in the dark-adapted individuals (Fig.6). In the training trials (days 8-14) the activity was increased after simultaneous illumination and food stimuli (solid lines in Figs.5 and 6) and lasted until the end of the observation period (broken lines). In the test trials, the food was presented after the end of the illumination period. In both cases individuals stopped responding to illumination alone except for day 15 in the dark-adapted indivi-

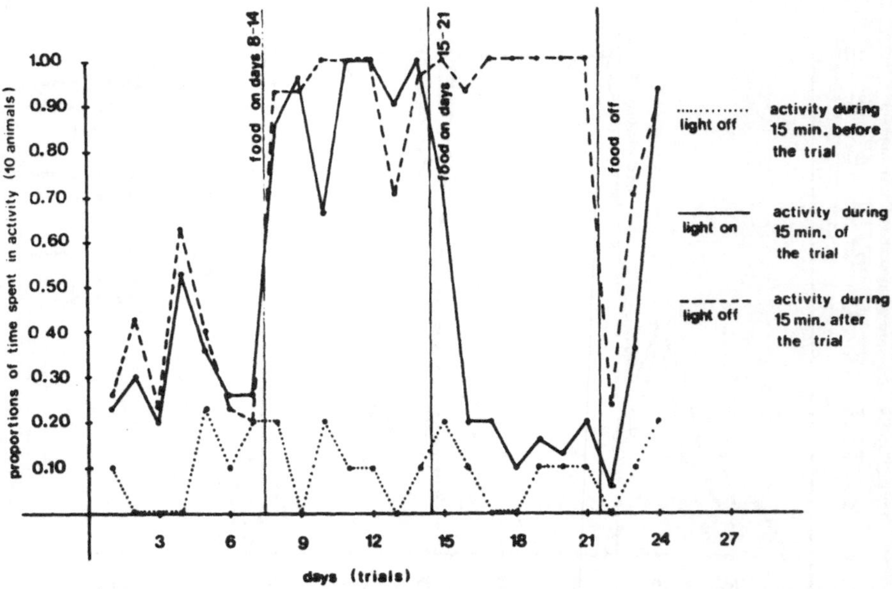

Figure 6. Activities of dark-adapted *Marthasterias glacialis* after exposure to light and food at 24-hour intervals (Valentinčič & Kranjc, unpublished).

duals. The chemical stimuli after the end of the light regimen change produced activity in all individuals. There was no food presented during days 22 to 24. The light-adapted individuals behaved like controls. The dark-adapted individuals increased in response to illumination in trials 23 and 24 after an initial decrease in activities.

Individuals, adapted to dim (5 to 10 lx) light, were exposed to a strong beam of light in the next experiment. Illumination as high as 15 000 lx (tungsten bulb) were employed. An intertrial period of two hours was used during the control experiments 1-7 (Fig.7). Several individuals remained inactive while some started moving as a result of illumination alone. Their activity decreased during the seven control trials, possibly indicating habituation. Nearly all the individuals in the training trials were active for the entire ten minutes after simultaneous presentation of light and the key stimulus for food (L-proline). Three successive training trials and a test were done with each animal on a single day. Light was presented alone in test trials numbers 13, 17, 21 and 25. The form of diagrams illustrating the activity of animals in a single test is not different from the initial controls.

The results of Landenberger on *Pisaster giganteus* were not found in our work with *Marthasterias glacialis.* Further work with dark-adapted animals might reveal interesting phenomena concerning learning or changes in responses to light.

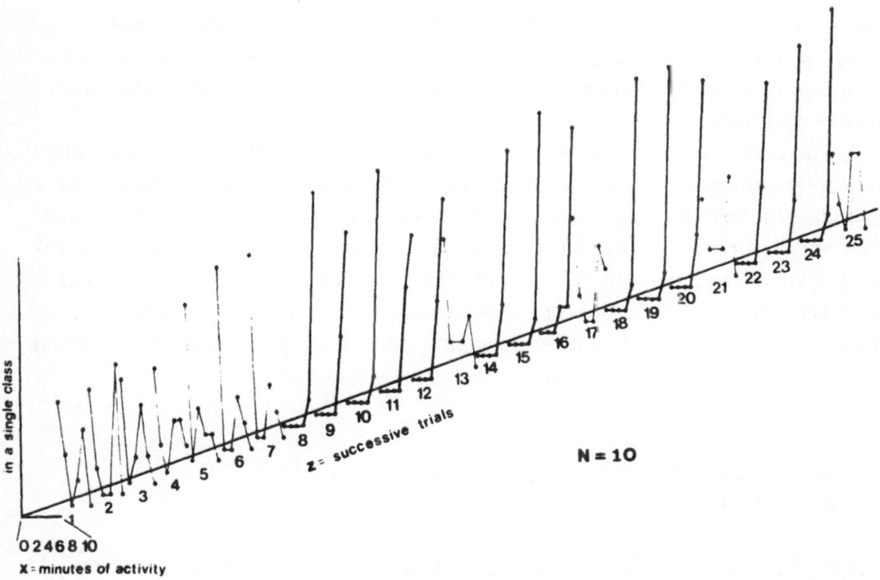

Figure 7. A three dimensional representation of the responses of *Marthasterias glacialis* to a strong beam of light and to a simultaneous presentation of light and food (L-proline) stimuli. The results of the control experiments are indicated by dots and thin lines (trials 1-7). The training trials are indicated by thick lines. Trials 13, 17, 21 and 25 indicate tests (Valentinčič & Kranjc, unpublished).

7 INGESTIVE CONDITIONING (CHEMOSENSORY CONDITIONING) IN ASTEROIDS

Landenberger (1968), Castilla (1972), and Ormond *et al.* (1976) worked on ingestive conditioning in asteroids, following the ideas of Thorpe (1938) and Wood (1968). The idea of ingestive conditioning is basically the same as that of the search image suggested first by Von Uexküll (1934) and developed by Tinbergen (1960). The term olfactory conditioning, as used by Thorpe in insects, probably does not have the same precise meaning when applied to aquatic animals. Chemosensory conditioning would describe the same phenomenon properly. Ingestive conditioning (chemosensory conditioning) might be the mechanism by which the search image is formed.

Landenberger (1968) found that *Pisaster giganteus* preferred mussels over turban snails *(Tegula funebralis).* A reduced preference for mussels was shown in the asteroids exposed to turban snails for three months. The searching behavior was turned back to mussels after only one week of exposure to them. The term search image correctly describes the experimental situation. Landenberger claimed that not only did the preference change, but that *Pisaster giganteus* moved to the bottom where the mussels were situated. This increased the consumption of mussels in comparison with the snails which occurred on the walls of the containers.

In the case of chemosensory conditioning, chemical stimuli from a potential prey were constantly present in the environment of the animal. This is not the case with the search image based on vision. The bird is capable of detecting its prey only intermitently when finding it. Thus long-term memory was not necessarily involved in Landenberger's experiments. The experiments of Ormond *et al.* (1976) are very suggestive for long-term memory. The preference of *Acanthaster planci* was turned to *Fungia fungites* and a clear preference for this coral was detected even after two months' diet on *Acropora* sp.

8 ASSOCIATIVE LEARNING IN *MARTHASTERIAS GLACIALIS*

Our experimental approach to associative learning problems in *Marthasterias glacialis* was based on the Tinbergen's (1951) view: 'Learning is a secondary modification of innate mechanisms'. A search for behavioral, electrophysiological, and chemical correlates of learning would be premature if the corresponding backgrounds of innate behaviors are not well understood. Feeding behaviors released by key stimuli for feeding and escape responses to a very strong stimuli are parts of the normal behavioral inventory of any animal species. The feeding behavior in *Marthasterias glacialis* was elicited at a distance by food, food homogenates, bacteriological peptone, L-cysteine, acetylcholine iodide, L-proline lactic acid, betaine, and several neutral amino acids (Valentinčič 1973, 1975, 1980a,b,c). Cotton balls soaked with a solution of key chemical stimuli for the feeding response release the entire sequence of feeding motions:

Figure 8. An arm tip in a resting *Marthasterias glacialis*.

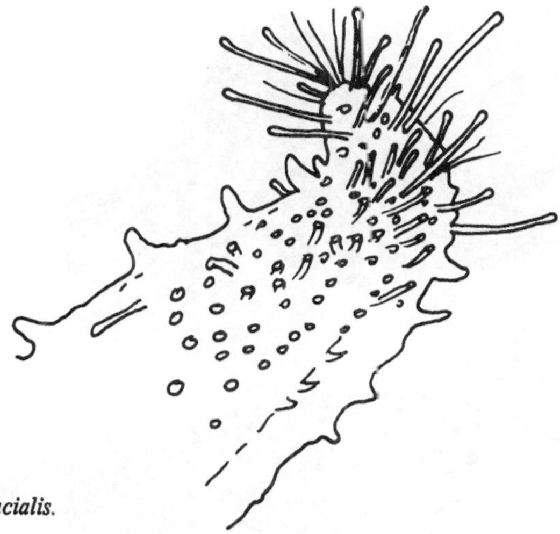

Figure 9. The orienting response of the arm tip in *Marthasterias glacialis.*

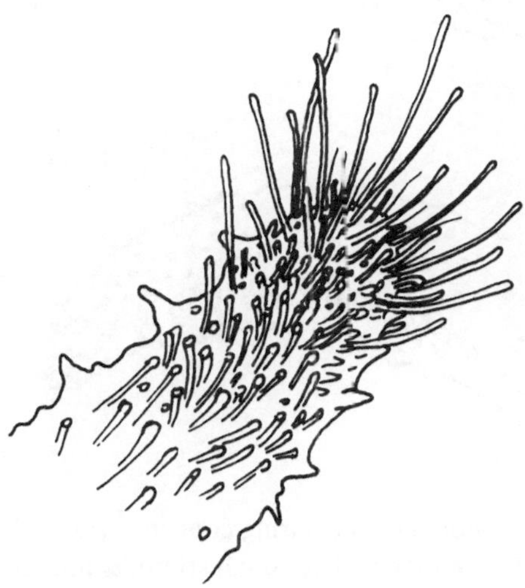

Figure 10. The leading arm during the appetitive behavior of *Marthasterias glacialis.*

– Orienting response (Figs.8 and 9): the tips of the arms lift up (the arm-curling response) and the tube feet situated there extend and show clinotactic movements.

– Appetitive behavior (Fig.10): individuals start to move their arms and crawl around in a search for higher concentrations of the stimulating chemicals.

– Consumatory ingestive behavior (Fig.11): after the tube feet touch the food model or food, the asteroid brings the model into the central position under the mouth – assumes the feeding posture and the stomach is pressed against the cotton.

Two distinct stimuli are available to the asteroids in the feeding response to the food models. The key stimulus for the feeding response releases the feeding

Figure 11. The consummatory feeding posture in *Marthasterias glacialis.*

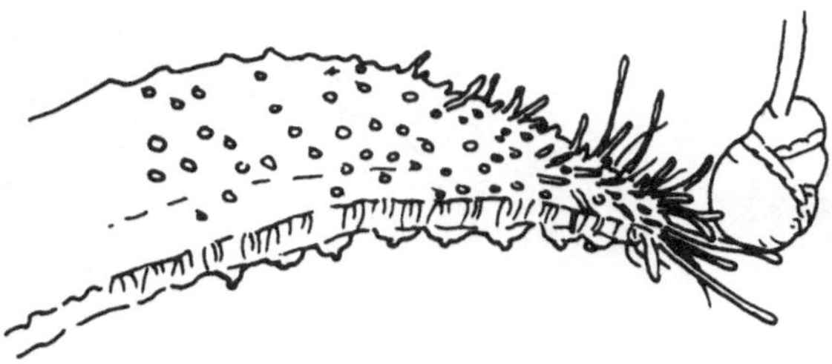

Figure 12. The dominant arm of an advancing *Marthasterias glacialis* and the 'electric' food-model.

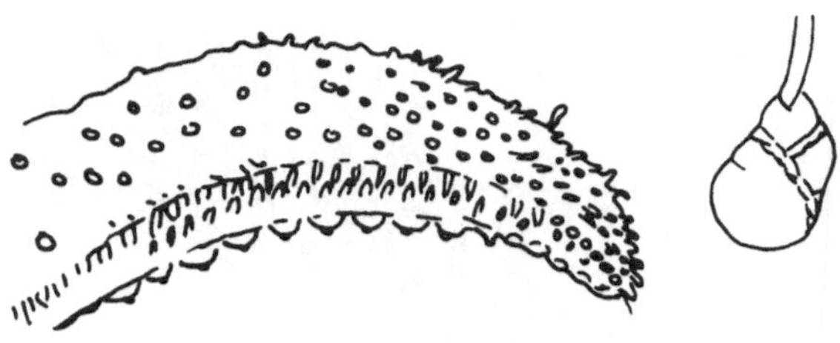

Figure 13. The escape response of the arm tip of *Marthasterias glacialis* to the 'electric' food-model.

behavior. The increasing concentrations of the chemical enables the asteroid to find the model. The touch stimulus provided by the surface structure of the food model does not produce the feeding behavior. Cotton balls identical to the food models are ignored even if the chemical stimulus is simultaneously present in aquarium water. The surface structure of the food model provides a stimulus which has no innate releasing properties.

The escape response in *Marthasterias glacialis* might, for instance, be elicited by a touch by echinoid pedicellariae and by weak electric shocks. It consists of tube-feet retraction, shortening of the affected arm, and escape of the asteroid in the direction opposite to the noxious stimulus (Figs.12 and 13).

The food models were modified by inplanting electrodes in the cotton to obtain the 'electric' food-model (Valentinčič 1978). The electric food-model was placed in front of the asteroid. Approximately 6 seconds after the tube feet had touched the model, a weak electric shock was delivered. Thus approach-

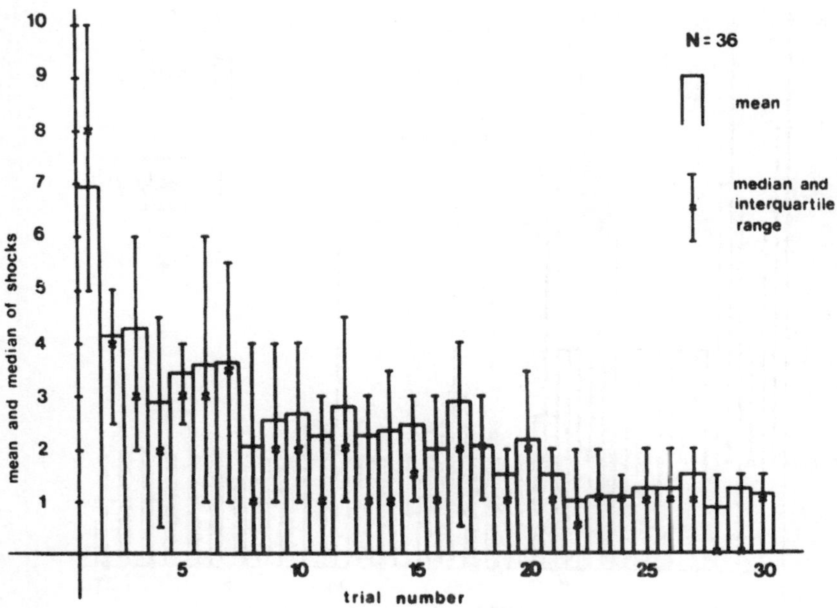

Figure 14. The number of punishments (electric shocks) necessary to produce a single escape or avoidance response in *Marthasterias glacialis* maintained under conditions of controlled starvation (from Valentinčič 1980).

avoidance conflict was provoked. The chemical stimuli emanating from the model were releasing the consumatory phase of the feeding behavior while the electric shocks elicited the escape response. At the beginning, several shocks were necessary to produce a single complete escape response. After the escape response declined, the animals frequently reattacked the models.

There is a decrease in the number of shocks necessary to produce the escape behavior response in individuals maintained under conditions of controlled starvation for a period of six months (Fig.14) and in individuals subjected to learning within two months after collection (Fig.15). The first avoidance response ordinarily occurred after 2 to 10 trials. The asteroids avoided the electric shocks by avoiding the food models – the escape response was elicited by the tactile information provided by the surface of the cotton. The median number of shocks necessary was smaller than its corresponding mean. Only a few individuals of the total experimental population had to be punished with a large number of shocks for a single avoidance to occur in later trials. Most individuals either avoided the food model or a single punishment produced the escape behavior.

The initial decrease in vigor of the consumatory feeding response which occurred in both cases after one to three successive trials could be a consequence

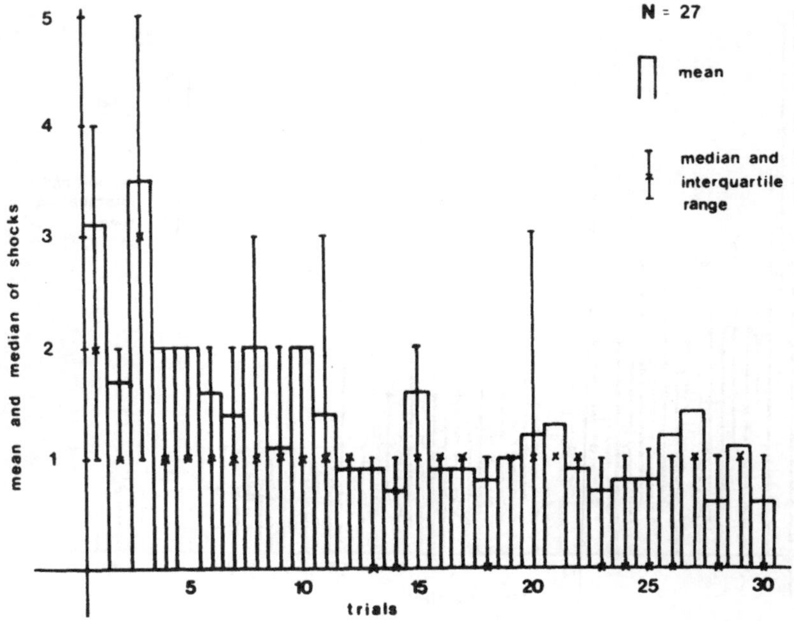

Figure 15. Number of punishments (electric shocks) necessary to produce a single escape or avoidance response in recently collected *Marthasterias glacialis* (Valentinčič & Lebez, unpublished).

of arousal for the escape behavior. The mean number of shocks for a single escape response decreased further after 7 to 12 trials. After more than 20 trials the mean values became low and the median values frequently were zero. The initial increase in the number of avoidance responses by individuals under controlled starvation that occurred during the first 11 to 14 trials was followed by a temporary decrease for the next four to five trials (Fig.16). This is a loss of learned response (Valentinčič 1978). The synchronous loss of the learned response in the entire experimental population probably reflects equal motivation in starved individuals. After approximately 25 trials, half of the individuals showed avoidance responses to the food models on each repetition.

An example of results obtained with a single individual is presented in Fig.17. The avoidance response first appeared after ten trials. Later it interchanged with intense feeding behavior several times during 115 successive presentations. Thus several losses of learned behavior seem to occur. There seems to be a kind of 'flip-flop' mechanism between feeding behavior and the avoidance response which controls the ensuing behavior. The probability for the avoidance response increases in successive trials, but feeding behavior may reappear at irregular intervals at its full strength. Two types of nervous connections appear to exist in the central nervous system of the asteroid: a feeding behavior network and

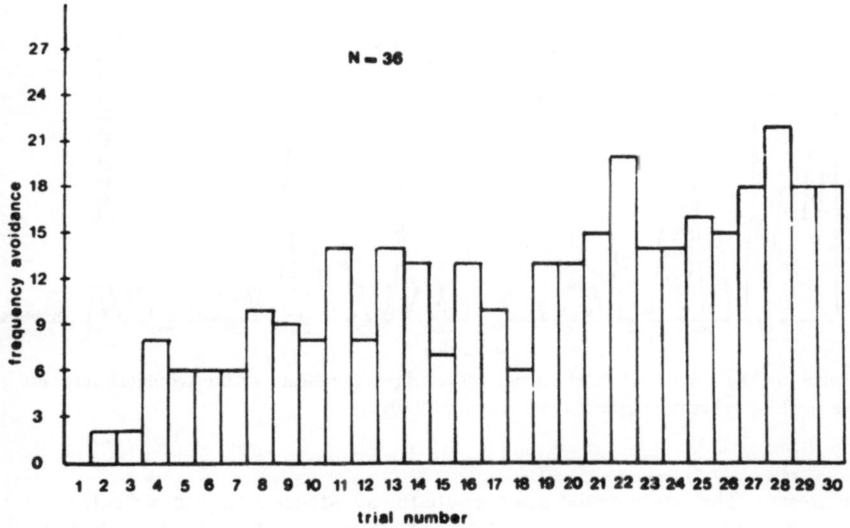

Figure 16. Avoidance responses to the 'electric' food-models in successive trials for 36 *Marthasterias glacialis.* (Valentinčič 1980).

an escape behavior network. The touch stimulus with the food model was associated by learning with the escape response, and the avoidance behavior occurred. Hunger pushed the 'flip-flop' mechanisms into the feeding mode, while punishments turned it to avoidance mode.

The validity of this hypothesis was tested within a series of experiments with a group of 16 individuals (Valentinčič & Lebez, unpublished). During 30 successive trials with the 'electric' food model containing L-cysteine, the number of avoidance responses increased in the first 13 trials, fluctuated for the next 10 trials, and then stabilized at a certain level in later presentations (Fig.18).

'Control electric-models' with no chemical stimulus in the cotton or in the sea water were presented to asteroids during their spontaneous activity a week later. Avoidance response occurred in half of the cases. This indicated either a long-term memory for the avoidance response or that the escape response is a part of an innate behavioral inventory.

In the next series of trials with the same individuals, the feeding behavior was released by 10^{-6} M solution of L-cysteine. The 'control electric-models' were soaked in aquarium water and presented in front of advancing individuals. The asteroids performed the avoidance response in more than half of the cases. The models were offered to the asteroids during the appetitive feeding behavior, but no consumatory feeding behavior occurred to the aquarium water-soaked models. A gradient of the chemical stimulus is thus necessary for a feeding posture to occur. The two regression lines (Fig.18) have a weak slope indicating further learning during the trials even if no key stimuli for the feeding response

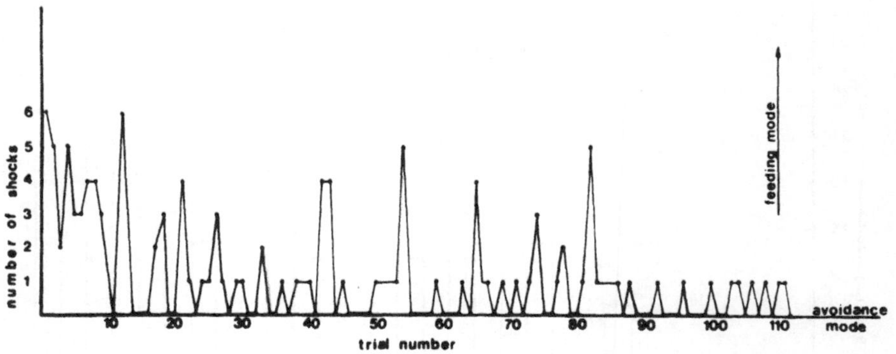

Figure 17. The number of shocks necessary to trigger the escape or the avoidance response in a single *Marthasterias glacialis* (Valentinčič 1980).

are present. The asteroids did not execute the avoidance response perfectly after 60 or 90 trials. It occurred in 75 to 100% of the individuals in each trial. The asteroids probably were not capable of recognizing the surface of the model, on each repetition. It is important to point out that a single shock was usually enough in the trials using the 'control food-models' and the escape responses followed. There was no feeding response to the food models and thus there was no approach-avoidance conflict in the last two experimental situations.

9 INNATE BEHAVIOR, LEARNING, AND NERVOUS FUNCTIONS IN ASTEROIDS

The nervous system of asteroids includes radial nerve cords containing ectoneural and hyponeural tissue, a circumoral nerve ring connecting the cords, and the entoneural nerve plexus. Discrete 'serially' arranged neuropiles occur within the radial nerve cords (Cobb 1970), indicating segmental organization of the radial nerve cords. The segmental organization of the radial nerve cords has been demonstrated conclusively in *Ophiura texturata* (Cobb & Stubbs 1981). Nervous cell bodies in the radial nerve cords occur within the segmental ganglia. The nerve fibers run from one segment to the other and no continuous giant fibers were detected in any part of the nerve cord.

The tube foot-ampulla system is 'dually' innervated. Cobb (1967) found that the structure Smith (1950) called the ribbon axon is composed of fine muscle tails. The muscle tails form 'synaptoid contacts' across the basement membrane with the centrally derived motor neurons in the tube foot nerve. The muscle tails occur in the tube foot muscles (Cobb 1970). Raymond (1980) confirmed the presence of a small number of muscle tails in the proximal region

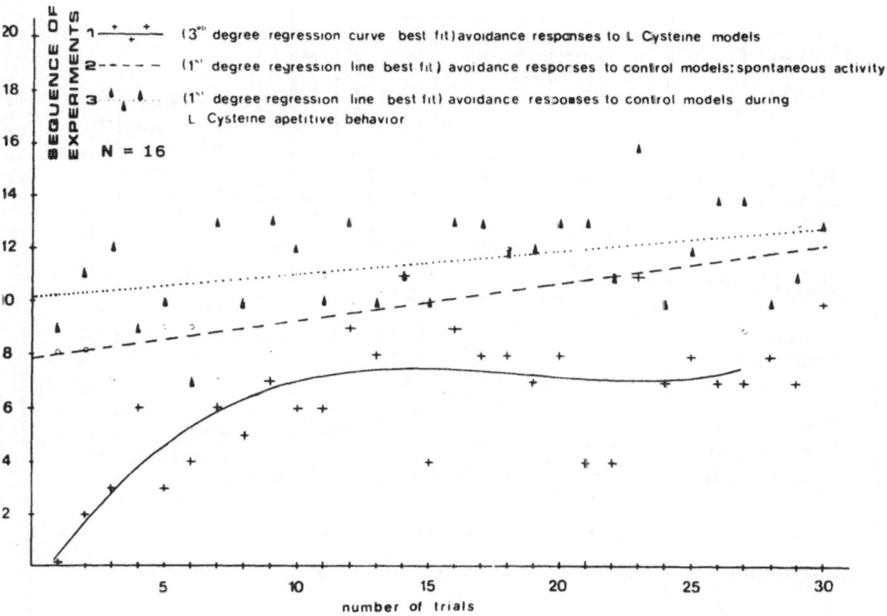

Figure 18. The avoidance responses of a group of 16 *Marthasterias glacialis* in three successive experiments with the same individuals. The interval between experiments was 3 to 6 days (Valentinčič & Lebez, unpublished).

of the stem retractor muscles in echinoids. The second type of innervation is provided by axons containing large granular vesicles (Cobb 1978) supplying the peripheral control of the tube feet musculature (Raymond 1980). The coordinated activities of the tube-feet would thus be controlled via the muscle tails.

The structure of sensory systems in asteroids is little known. Most authors claim massive sensory input from the epithelial cells into the basiepithelial plexus. Weber & Grossman (1977) showed that sensory functions in echinoids might be limited to the ciliated epithelial cells and only to a limited number of other epithelial cells which send their processes into the basiepithelial plexus.

We can postulate functions and nervous elements which would make the observed behaviors possible. Inactivity, feeding as a motivated behavior, the escape and/or the avoidance response, the fright response, and the locomotion appearing in many different behaviors should be associated with specific nervous activities. All observed activities except for locomotion exclude the other behaviors. The same motor neurons and the same muscles are involved in all tube-feet motions. The information for activity and even for inactivity should be centrally propagated and specific central nervous states should be formed for each particular type of behavior. The whole set of the radial nerve cords

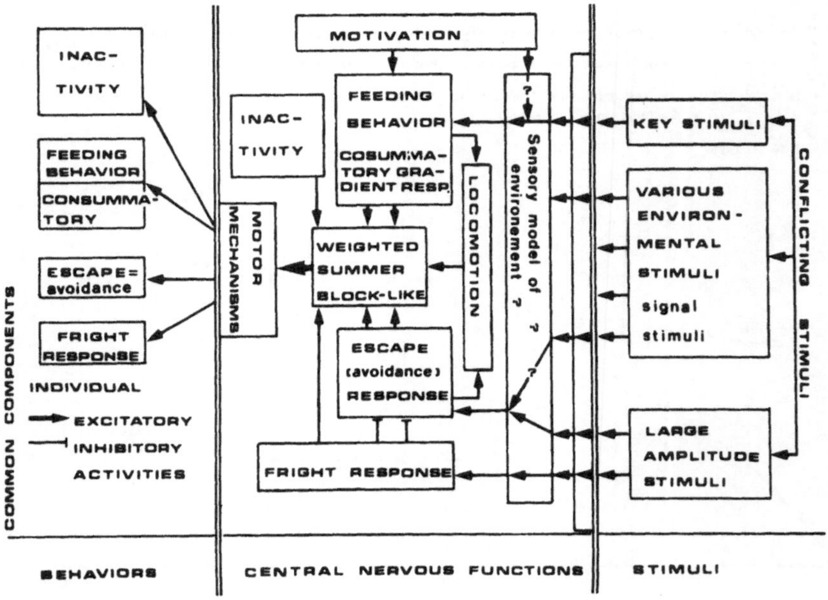

Figure 19. A simple neuronal model enabling the consumatory phase of the feeding response and the escape or avoidance response in *Marthasterias glacialis*. Other simple behaviors are also included.

and the interconnecting circumoral ring should be interpreted as a central nervous system under these assumptions. The fact that each tube-foot or each pedicellaria may act independently of the rest of the organism and show reflex activities does not decrease the necessity for the major role of segmental ganglia and the whole radial nerve cords in coordinated activities.

The reflex responses of single tube feet would result in confusion even in cases of directional stimuli. Sensory input should create a state in the distal segments of the radial nerve cord which enables the coordinated movements of all tube feet close to the arm tip. Thus a function of a leading arm may be established. It is possible that the rest of the body follows the leading arm(s) for mechanical reasons. Alternatively the nervous system state is propagated from 'ganglion' to 'ganglion' and the tube feet are coordinated by interplay between the central nervous state and reflex responses to the environmental stimuli. The tube feet retraction in the escape response or in the avoidance response is always propagated far beyond the stimulated area. The 'central escape state' thus travels beyond the stimulated 'ganglia'. The most extreme case is the fright response. Grasping of a single arm of the asteroid in an attempt to detach it from the substrate produces a very typical behavioral change in the whole organism. Most of the tube feet strongly grip the substrate after only a few seconds and the asteroid can be pulled away only by breaking many of the tube feet. Only a specific central nervous state might induce such simultaneous long-lasting actions of all the tube feet muscles.

My primary concern in the development of a neuronal model (Fig.19) is the appearance of the consumatory phase of the feeding response and the escape or avoidance responses to the 'electric' food-models. In the first trials several shocks were necessary to produce an escape response to the food model. With successive trials the number of shocks to produce escape usually decreases as if the system were behaving like a weighted summer-block type from the continuous system modeling programs for computers. Such a comparison has no implications for structures and physiology of tissues involved in behavior and the learning phenomenon itself, but some functional resemblances are more than suggestive. A description for the two input-weighted summer blocks can be used in considering only the consumatory feeding response and the escape or avoidance response:

$$e_o = P_1 e_1 + P_2 e_2$$

e_o: occurring behavior,
e_1: consumatory phase of the feeding behavior,
e_2: the escape response,
P_1: 'readiness' for the consumatory phase of the feeding behavior,
P_2: 'readiness' for the escape response.

The punishments increased the readiness for the escape response with the successive training trials, while hunger and feeding stimuli increased the readiness for the feeding response. The avoidance response to the signal stimulus occurred when readiness for the escape response reached a level above that for the consumatory phase of the feeding behavior.

The introduction of logics known from programming systems as digital-analog simulators might help in understanding the functions which should be accomplished by structures and physiological mechanisms yet unknown to us. The black box indicating the sensory model of the environment is based on Sokolov's (1960) 'system for formation of model'. The stimulus which has no innate releasing properties might become a signal stimulus for the avoidance response. The association of the 'painful' stimuli with the signal stimuli should thus be accomplished in structures other than those enabling the motor actions of the avoidance response. The fact that the long-term memory has not been conclusively demonstrated in asteroids implies that, at least for the short-term memory, the association of the large amplitude stimuli and the signal stimuli is not absolutely necessary. A very unspecific stimulus might produce the escape response when the readiness for it is high.

It is nearly impossible to classify the avoidance learning in *Marthasterias glacialis* into any of the learning types as known in vertebrates. Learning is a change in behavior produced by experience and it is not attributable to growth, fatigue, and sensory adaptation. With this in mind, the decrease in mean number of shocks to produce the escape response and the appearance of the avoidance response can be considered a consequence of learning. There is no basis to attribute the 'seeming' long-term retention of the avoidance response to the long-term memory phenomena. An experimental demonstration of long-term memory in the asteroid would support the associative nature of the avoidance response. On the contrary, if only a notion of the short-term memory is applicable, either sensitisation, or associative learning might be the process involved. Further experiments are needed to clarify these questions.

10 CONCLUSIONS

Innate responses to external stimuli are described in resting and in locomoting asteroids. An experimental manipulation determines the behavior of *Marthasterias glacialis.* The direction of movements is maintained if the subordinate arms are manipulated and can be affected by grasping the dominant ones. The old view of Jennings of the persistence of movement is thus confirmed. Many of the works on rheo- and chemo-sensitivity in asteroids conducted in flow-tanks and in Y-mazes could be reinterpreted.

The clear-cut results of Landenberger regarding light and food stimuli in *Pisaster giganteus* could not be repeated in our experiments with *Marthasterias glacialis.* On the other hand, reports of ingestive (chemosensory) conditioning (search image) are sufficiently supported in asteroids.

Investigations of the avoidance response to the 'electric' food-model illustrate changes in behavior in an approach-avoidance conflict. The punishments (electric shocks) increase the 'readiness' of individual *Marthasterias glacialis* for escape and for the avoidance response. A kind of short-term memory seems to build up by an increase in the 'readiness' for the escape response in comparison with the 'readiness' for the consumatory phase of the feeding behavior. A tentative neuronal model considering both behaviors was constructed. Two possibilities are open. The avoidance responses can be triggered by the surface structure of the model serving either as an inadequate stimulus releasing the escape response, or as a stimulus associated by learning with the escape response.

Short-term memory seems to be documented while long-term changes in behavior originating in learning remain to be discovered in asteroids.

11 ACKNOWLEDGEMENTS

I wish to thank Professor Matija Gogala for his constructive criticism and support. Particular thanks should be given to my friend, mathematician Andrej Blejec for many stimulating discussions and suggestions. I would also like to acknowledge, for improvements of my English, Professor John M.Lawrence. The technical help was provided by Mario Peruzzi, Malči Blaževič, Danilo Musar, and Emerik Mezgolits.

12 REFERENCES

Applewhite, P.B. 1979. Learning in Protozoa. In: M.Lewandowsky & S.N.Hunter (eds), *Biochemistry and Physiology of Protozoa* (2nd ed.), Vol.1, pp.341-355. Academic Press, New York.

Barness, D.J., R.W.Brauer & M.R.Jordan 1970. Locomotory response of *Acanthaster planci* to various species of coral. *Nature, Lond.* 228:342-344.

Bastock, M., D.Morris & M.Moynihan 1953. Some comments on conflict and thwarting in animals. *Behaviour.* 6:66-84.

Binyon, J. 1972. *Physiology of Echinoderms.* Pergamon Press, Oxford.

Brauer, R.W. & M.R.Jordan 1970. Triggering of the stomach eversion reflex of *Acanthaster planci* by coral extracts. *Nature, Lond.* 228:344-346.

Castilla, J.C. 1971. Responses to light of *Asterias rubens.* In: D.J.Crisp (ed.), *Fourth Europ. mar. biol. Symp.*, pp.495-511. Cambridge University Press.

Castilla, J.C. 1972a. Responses of *Asterias rubens* to bivalve prey in a Y-maze. *Mar. Biol.* 12:222-228.

Castilla, J.C. 1972b. Avoidance behavior of *Asterias rubens* to extracts of *Mytilus edulis*, solutions of bacteriological peptone, and selected amino acids. *Mar. Biol.* 15:236-245.

Castilla, J.C. & D.J.Crisp 1970. Responses of *Asterias rubens* to olfactory stimuli. *J. mar. biol. Ass. UK*, 50:829-847.

Castilla, J.C. & D.J.Crisp 1973. Responses of *Asterias rubens* to water currents and their modification by certain environmental factors. *Neth. J. Sea Res.* 7:171-190.

Cobb, J.L.S. 1967. The innervation of the ampulla of the tube foot in the starfish *Astropecten irregularis. Proc. Roy. Soc.* (B)168:91-99.

Cobb, J.L.S. 1970. The significance of the radial nerve cords in asteroids and echinoids. *Z. Zellf.* 108:457-474.

Cobb, J.L.S. 1978. An ultrastructural study of the dermal papulae of the starfish *Asterias rubens* with special reference to innervation of the muscles. *Cell Tissue Res.* 187:515-523.

Cobb, J.L.S. & T.R.Stubbs 1981. The giant neuron system in ophiurids. *Cell Tissue Res.* 219:197-207.

Diebschlag, E. 1938. Ganzheitliches Verhalten und Lernen bei Echinodermen. *Z. vergl. Physiol.* 25:612-654.

Feder, H. & A.M.Christensen 1966. Aspects of asteroid biology. In: R.A.Boolootian (ed.), *Physiology of Echinodermata*, pp.87-127. Interscience Publ., New York.

Fenchel, T. 1965. Feeding biology of the sea-star *Luidia sarsi* Düben & Koren. *Ophelia* 2: 223-236.

Hyman, L.H. 1955. *The Invertebrates. Vol.IV. Echinodermata.* McGraw Hill, New York.

Jennings, H.S. 1906. *Behavior of Lower Organisms.* Columbia Univ. Press, New York.

Jennings, H.S. 1907. Behavior of the starfish *Asterias forreri* de Loriol. *Univ. Calif. Publ. Zool.* 4:53-158.

Kalmus, H. 1929. Versuche über die Bewegungen der Seesterne, besonders von *Asterina gibbosa. Z. vergl. Physiol.* 9:703-733.

Kerkut, G.A. 1954. The mechanisms of coordination of the starfish tube feet. *Behaviour,* 6:206-232.

Kjerskog-Agersborg, H.P. 1918. Bilateral tendencies and habits in the twenty-rayed starfish *Pycnopodia helianthoides* (Stimpson). *Biol. Bull.* 35:232-254.
Kjerskog-Agersborg, H.P. 1922. The relation of the madreporite to the physiological anterior end in the 20-rayed starfish *Pycnopodia helianthoides. Biol. Bull.* 42:202-216.
Landenberger, D.E. 1966. Learning in the Pacific starfish *Pisaster giganteus. Anim. Behav.* 14:414-418.
Landenberger, D.E. 1968. Studies on selective feeding in the Pacific starfish *Pisaster* in Southern California. *Ecology* 49:1062-1075.
Lorenz, K. 1935. Der Kumpan in der Umwelt des Vogels. *J.f. Ornith.* 83:137-413.
Mangold, E. 1908. Studien zu Physiologie des Nervensystem der Echinodermen. I. Die Füsschen der Seesterne und die Koordination ihrer Bewegungen. *Pflüg. Arch. ges. Physiol.* 122:315-360.
Massé, H. 1975. Ethologie alimentaire de *Astropecten aranciacus* L. In: H.Barnes (ed.), *Proc. 9th Europ. mar. biol. Symp.*, pp.343-355.
Millott, N. 1966. Coordination of spine movement in echinoids. In: R.A.Boolootian (ed.); *Physiology of Echinodermata*, pp.465-486. Interscience Publishers, New York.
Mori, S. & K.Mutatani 1952. Studies on the daily rhythmic activity of the starfish, *Astropecten polyacanthus* Müller et Troschel, and the accompanied physiological rhythms. *Publ. Seto mar. biol. Lab.* 2:213-225.
Ormond, R.F.G. & A.C.Campbell 1974. Formation and breakdown of *Acanthaster planci* aggregations in the Red Sea. *Proc. 2nd Int. Coral Reef Symp. Brisbane*, pp.595-619.
Ormond, R.F.G., N.J.Hanscomb & D.A.Beach 1976. Food selection and learning in the crown-of-thorns starfish *Acanthaster planci* (L.). *Mar. behav. Physiol.* 4:93-105.
Ormond, R.F.G., A.C.Campbell, S.H.Head, R.J.Moore, P.R.Rainbow & A.P.Saunders 1973. Formation and breakdown of aggregations of the crown-of-thorns starfish, *Acanthaster planci* (L.). *Nature, Lond.* 246:167-169.
Preyer, W. 1887. Uber die Bewegungen der Seesterne. *Mitt. zool. Sta. Neapel* 7:191-233.
Raymond, A.M. 1980. Innervation of echinoid tube feet. In: M.Jangoux (ed.), *Echinoderms: Present and Past*, pp.277-283. Balkema, Rotterdam.
Reese, E.S. 1966. The complex behavior of echinoderms. In: R.A.Boolootian (ed.), *Physiology of Echinodermata*, pp.157-218. Interscience Publishers, New York.
Romanes, G.J. 1883. Observations on the physiology of Echinodermata. *J. Linn. Soc. (Zool.)* 17:131-137.
Sloan, N.A. 1980. Aspects of feeding biology of asteroids. *Oceanogr. mar. Biol. ann. Rev.* 18:57-124.
Sloan, N.A. & A.C.Campbell 1982. Perception of food. In: M.Jangoux & J.M.Lawrence (eds), *Echinoderm nutrition*, pp. 3-23, Balkema, Rotterdam.
Smith, J.E. 1950. The motor nervous system of the starfish, *Astropecten irregularis* (Pennant) with special reference to the innervation of the tube feet and ampullae. *Phil. Trans. Roy. Soc.* (B)234:521-558.
Sokolov, E.N. 1960. Neuronal models and the orienting reflex. In: M.A.B.Brazier (ed.), *The Central Nervous System and Behavior*. Josiah Macy Jr. Foundation, New York.
Sokolov, V.A. 1961. Tactile conditioning in *Asterias rubens. Trud. Murmansk. morsk. biol. Inst.* 3:49-54 (in Russian).
Thorpe, W.H. 1938. Further experiments on the olfactory conditioning in a parasitic insect. The nature of conditioning process. *Proc. Roy. Soc.* (B):126:370-397.
Thorpe, W.H. 1956. *Learning and Instinct in Animals.* Methuen & Co., London.
Tinbergen, L. 1960. The natural control of insects in pinewoods: I. Factors influencing the intensity of predation by song birds. *Arch. néerl. Zool.* 13:265-345.
Tinbergen, N. 1951. *The study of Instinct.* Clarendon Press, Oxford.

Tortonese, E. 1950. Ricerche sperimentali e comparative sui movimenti degli asteroidi. *Arch. zool. Ital.* 35:217-250.

Uexküll, J.von 1934. *Streizüge durch die Umwelten von Tieren und Meschen.* Springer, Berlin. English translation: *Instinctive Behaviour* (1957), Ed. Schiller, C.H.Methuen, London.

Valentinčič, T. 1973. Food finding and stimuli to feeding in the sea star *Marthasterias glacialis. Neth. J. Sea Res.* 7:191-199.

Valentinčič, T. 1975. Amino-acid chemoreception and other releasing factors in the feeding response of the sea star, *Marthasterias glacialis* (L.). In: H.Barnes (ed.), *Proc. 9th Europ. mar. biol. Symp.* pp.275-286. Aberdeen University Press.

Valentinčič, T. 1978. Learning in the starfish, *Marthasterias glacialis.* In: D.J.McLusky & A.J.Berry (eds), *Physiology and Behaviour of Marine Organisms,* pp.303-309. Pergamon Press, Oxford.

Valentinčič, T. 1980a. Associative learning in the starfish *Marthasterias glacialis,* a simple model for the study of learning. In: M.Jangoux (ed.), *Echinoderms: Present and Past,* pp.337-341. Balkema, Rotterdam.

Valentinčič, T. 1980b. Structure-activity relationships of amino acids in *Marthasterias glacialis* chemoreception. In: M.Jangoux (ed.), *Echinoderms: Present and Past,* pp.331-335. Balkema, Rotterdam.

Valentinčič, T. 1980c. Structure-activity studies of amino acids and acetylcholine in chemoreception of the starfish, *Marthasterias glacialis.* In: M.Brzin, D.Sket & H.Bachelard (eds), *Synaptic Constituents in Health and Disease.* pp.683. Mladinska knjiga, Ljubljana & Pergamon Press, Oxford.

Ven, C.D. 1921. Sur la formation d'habitudes chez les astéries. *Archs Néerld. Physiol.* 6:163-178.

Weber, W. & M.Grosmann 1977. Ultrastructure of the basiepithelial nerve plexus of the sea urchin, *Centrostephanus longispinus. Cell Tissue Res.* 175:551-562.

Willows, A.O.D. & W.C.Corning 1975. The echinoderms. In: W.C.Corning, J.A.Dyal & A.O.D.Willows (eds), *Invertebrate Learning.* Vol.3, pp.103-135. Plenum Press, New York & London.

Whittle, K.J. & N.Blumer 1970a. A predator-prey relationship. Sea star-bivalves. The chemical basis of the response of *Asterias vulgaris* to *Crassostrea virginica.* A bioassay, its application and the partial purification of an active extract. Woods Hole Oceanographic Institution, Technical Report, Ref. No.70-20.

Whittle, K.J. & M.Blumer 1970b. Interaction between organisms and dissolved organic substances in the sea. Chemical attraction of the starfish *Asterias vulgaris* to oysters. In: D.W.Hood (ed.), *Organic Matter in natural Waters.* Inst. Mar. Sci. U. of Alaska, Pub. No.1, pp.495-507.

Wood, L.H. 1968. Physiological and ecological aspects of prey selection in the marine gastropod *Urosalpix cinerea* (Prosobranchia: Muricidae). *Malacologia* 6:267-320.

Yoshida, M. 1966, Photosensitivity. In: R.A.Boolootian (ed.), *Physiology of Echinodermata,* pp.435-464. Interscience Publishers, New York.

Zafiriou, O. 1972. Responses of *Asterias vulgaris* to chemical stimuli. *Mar. Biol.* 17:100-107.

Zafiriou, O., K.J.Whittle & M.Blumer 1972. Responses of *Asterias vulgaris* to bivalves and bivalve tissue extracts. *Mar. Biol.* 13(2):137-145.

FORM AND FUNCTION OF PEDICELLARIAE

ANDREW C. CAMPBELL
Department of Zoology & Comparative Physiology, Queen Mary College, London, UK

CONTENTS

1 INTRODUCTION

'If the echinoderms had not been discovered would they have been invented?'

The outer surface of an echinoderm is an interface between an animal and its environment of more than usual interest. Here we should expect to find structures which reflect the processes taking place at this interface. The external surfaces of most echinoderms have some features in common, e.g. tube feet, spines, and mucus-secreting glands, but asteroids and echinoids also have pedicellariae. Pedicellariae are small pincer-like organs with movable jaws known as valves. The most outstanding feature of pedicellariae is their independent reflex behavior mediated by their own sensory cells, neuropil and muscles. In many cases they give reflex responses even when isolated from the test and its basi-epithelial nerve plexus. In addition, stimuli applied to the test can bring about certain responses. Asteroid pedicellariae may be attached directly to the test or carried on stalks; in echinoids they are always carried on stalks which can be moved around rather like the spines. Although some asteroids lack pedicellariae altogether, many species bear them, on the aboral surface and bordering the ambulacra; they occur in all echinoids, often in vast numbers.

What are pedicellariae for? Why should there be so many? Most echinoderm specialists have views on their everyday role, but there are surprisingly few published accounts of their natural activity. This is largely due to their small size, number and variety, to say nothing of their capricious behavior, which makes them difficult subjects for study. Although pedicellariae have attracted a number of workers, several aspects of their microanatomy remain to be investigated. Their tissues are not easy to fix well, and decalcification poses further problems for ultrastructural analysis. Their behavior has so far been tested only by simple experiments as they are highly resistant to standard electrophysiological techniques.

2 DISCOVERY, EARLY HISTORY AND NOMENCLATURE

Pedicellarial research has progressed broadly in line with other anatomical and physiological investigations of echinoderms. Generally regular echinoids have received more attention than irregular echinoids or asteroids. Although asteroids and echinoids were quite well known by the ancient Greeks, and their spines recognised as protective structures (Thompson 1947), pedicellariae were unrecognised until 1788 when Müller described them as polyps parasitic on the echinoid *Echinus saxatilis (= Psammechinus miliaris).* He believed that there were three types, and named them generically as *Pedicellaria globifera, P. triphylla* and *P. tridens.* Valentin (1841) was more perceptive and recognised that the pedicellariae were part of the echinoid itself. His system of nomenclature was derived from observations on *Echinus esculentus* Linnaeus: 'gemmiforme', 'ophiocephale' and 'tridactyle'. He also used the term 'buccale' to describe the aggregation of ophiocephalous organs situated around the mouth of the echinoid. The nomenclature of these two authors, together with that of other pioneers and principal contributors is given in Table 1, which also includes the names of the minor types of echinoid pedicellariae. Despite Valentin's work, L.Agassiz published an account in the same year (1841) in which pedicellariae were described as young forms of the echinoid. This idea was eventually refuted by Gosse (1856) on the grounds that the population of pedicellariae on the test of a given individual remained constant throughout the year,

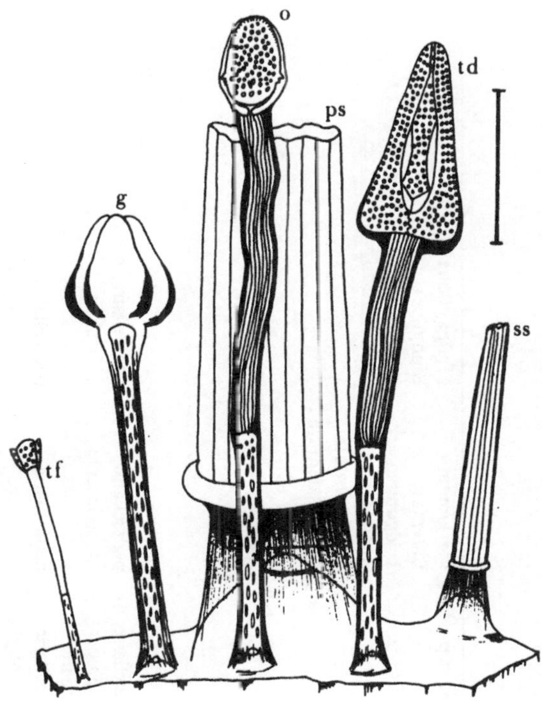

Figure 1. Diagram, from a drawing by Beddard (see Geddes & Beddard 1881), showing the principal test appendages of *Echinus sphaera (= E.esculatus)*; g: globiferous pedicellaria; o: ophiocephalous pedicellaria; ps: primary spine; ss: secondary spine; td: tridentate pedicellaria; tf: trifoliate pedicellaria. Scale bar 1 mm.

so that they were unlikely to be 'infant echini'. In this respect Gosse had touched on an interesting point which will be taken up later. While Valentin noted that there were some very small pedicellariae among his 'ophiocephale', it was Perrier (1869) who described these as a separate category: trifolié.

The existence of two alternative schemes for classifying pedicellariae has led to some confusion in the echinoid literature. Mortensen (1903) held that Müller's (1788) terms should stand on the basis of priority; nevertheless he used 'ophiocephalous' in the sense of Valentin. This precedent was followed by Hyman (1955), and while there have been some departures from it, this scheme is now in general usage. It should be noted (Table 1) that the name triphyllous has acquired a meaning different from that intended originally by Müller, and that it now describes a category quite separate from the ophiocepahlous. Figure 1 shows the four types of echinoid pedicellariae as illustrated by Beddard (see Geddes & Beddard 1883). It should be noted also that the number of valves per pedicellaria is constant in most species, but there may be occasional exceptions to this, a fact appreciated long ago by Stewart (1879).

The early work on asteroid pedicellariae was reviewed by Perrier in his comprehensive thesis on pedicellariae of asteroids and echinoids (1869). Sharpey (1836) gave the earliest written description of asteroid pedicellariae. Müller & Troschel (1842) in *System der Asteriden* figured these organs, as did

Table 1. Synonymy of pedicellariae of echinoids.

Hyman 1955 and most subsequent authors	**Globiferous**	**'Claviform'** derived from globiferous by loss of head and retention of stalk glands	**Claviform** derived from ophiocephalous by growth of stalk glands: head loss	**Ophiocephalous**	**(Tri)dentate** Bidentate Tridentate Rostrate Quadridentate Quinquedentate	**(Tri)phyllous = Foliate** Biphyllous Triphyllous	**Dactylous**
Mortensen 1903			claviform				
Hamann 1887	gemmiformen	globiferen					
Döderlein 1887, 1906		claviform					
Foetinger 1881		claviform					
Thomson 1874							dactylous
Perrier 1869	pedicellaire gemmiforme			pedicellaire ophiocephale	pedicellaire tridactyle	pedicellaire trifolie	
Gosse 1856	pedicellaria globifera			pedicellaria tryphylla	pedicellaria tridens	pedicellaria stereophylla	
Valentin 1841	pedicellaire gemmiforme			pedicellaire ophiocephale pedicellaire buccale	pedicellaire tridactyle		
Müller 1788	*Pedicellaria globifera*			*Pedicellaria triphylla*	*Pedicellaria tridens*	*Pedicellaria triphylla*	

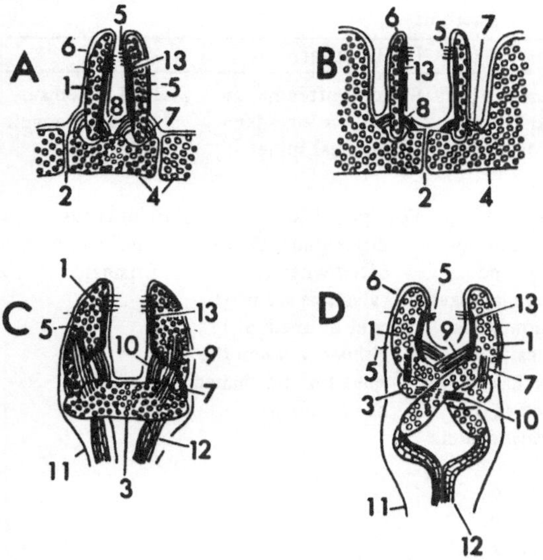

Figure 2. Diagram of principal types of asteroid pedicellariae to relate sensory areas, nerve, muscle, connective tissue and skeletal ossicles; A: sessile pedicellaria; B: alveolar pedicellaria; C: straight pedunculate pedicellaria; D: crossed pedunculate pedicellaria. 1: valve ossicle; 2: suture between test ossicles, 3: basal ossicle, 4: test ossicle, 5: tactile receptors, 6: epithelium, 7: abductor muscles, 8: adductor muscles, 9: long adductor muscles, 10: short adductor muscles, 11: stem, 12: connective tissue ligament, 13: valve nerve (C and D compiled from Smith 1937).

Duvernoy (1849) writing principally on echinoid morphology. Perrier (1869) mentioned that De Quatrefages, Dujardin, Hupe & Norman all discussed asteroid pedicellariae, but he did not give detailed references. By 1869 it was clear that *Asterias (= Asteracanthion)* had two types of pedicellariae. Work on asteroid pedicellariae has never been as popular as that on echinoids, but Perrier (1869) set the scene for the recognition of different forms. We now recognise three general types of asteroid pedicellariae: pedunculate, sessile and alveolar. The first type may have the valves crossed, like scissors, or straight, like forceps (Fig.2). The distribution and categories of pedicellariae in different orders of asteroids and echinoids is given in Tables 2 and 3.

3 ANALYSIS OF THE EFFECTOR ORGAN POTENTIAL OF PEDICELLARIAE

3.1 *General*

The suggestions of early workers as to the role of pedicellariae were based on observation and inference (Table 4). These suggestions acted as the germ for

Table 2. Distribution of pedicellariae in asteroids.

Sessile	Alveolar	Pedunculate	
Formed from adjacent spines: if stalked, very simple	Formed from adjacent spines sunk in an alveolus or pit: not stalked	Stalked, often marginal, e.g. on adambulacral spines	Stalked, often surrounding abactinal spines
Some paxillosids; Some platyasterids; *Straight sessile spiniform:* formed from long spines; *Pectinate,* formed from short curved spines often arranged in 2 rows; often meeting over a suture; *Fasciculate,* formed from a cluster of spines borne on one ossicle	Valvatids; *Bivalved* when 2 spines become wide and valve-like; may have powerful muscles sunk in test*; *Spatulate* long valves broadened at the tips; *Excavate:* long valves like tongs fitting into recesses of the alveolus when opened	Forcipulatids; Zorocallids; *Duck-billed* when the valve tips are bent over or unequal; *Felipedal,* when the valve tips are shaped like a cat's paw	Forcipulatids Euclasterids Brisingids

It is difficult to link the distribution of pedicellariae with habitat, but in general burrowing forms either lack pedicellariae or do not show such well developed ones as those species living on hard substrates: in asteroids pedicellarial complexity appears phylogenetic, the most highly evolved starfish bearing only crossed pedicellariae.

* See Haubold (1933)

Note: Some starfish which as adults lack pedicellariae bear them on larvae; i.e. some astropectinids (Downey, personal communication).

theories on pedicellariae published subsequently, namely that these organs are used for gripping and seizing objects in connection with waste removal, cleaning, clearing of sediments, locomotion, feeding and defence. The difficulty in analysing function is that pedicellariae are diverse in form, capricious in behavior, and have not lent themselves to detailed electrophysiological study. In short they are like the animals which bear them. There has been a strong inclination to regard pedicellariae as discrete effector organs of adult echinoids, divorced from spines as well as from each other in terms of control and function. Because of their usefulness as taxonomic indicators, greatly exploited by Mortensen (1903, 1928-1951), they are regarded as constant key characters of given species. That the pedicellariae might be in a state of flux reflecting the dynamic circumstances surrounding a particular individual is a new concept. This implication of constant form should not conceal the fact that the population of pedicellariae on an individual may be changing.

Since pedicellariae are effector organs, it would seem more wortwhile to relate their behavior to their intricately evolved structure and distribution in time and space to the environment and life of the individual that bears them, rather than to catalogue their conspicuous features. Pedicellariae are well adapted for seizing objects which touch or pass close to the body as well as for holding on to things they catch. This has been accepted by all workers in the field since Valentin (1841). These adaptations relate to four main areas:

1) Receptor cells whose form and distribution signal the presence of objects worthy of attack.

2) A nervous system controlling all opening and closing responses. This must integrate opening, closing and the venom extrusion responses in globiferous pedicellariae.

3) A system of muscles including abductors, adductors and flexors, as well as the venom extrusion muscles of globiferous pedicellariae.

4) A suitable skeleton for forming either a forcep-like or a pincer-like mechanism.

In all cases except the globiferous pedicellariae of echinoids, the overall shape of the organs reflects closely the form of the jaw ossicles themselves.

3.2 *Receptors*

Echinoderms lack complex receptor organs (Pentreath & Cobb 1972), but many species respond to touch, chemicals, and light. They may also be sensitive to water-born vibrations and pressure. In order to respond to fine particles such as silt and encrusting larvae, very sensitive tactile receptors are required.

3.2.1 *Echinoids*

Campbell & Laverack (1968), Chia (1969a,b) and Campbell (1974, 1976) have studied the effects of tactile and various chemical stimulation of echinoid pedicellariae valves. The areas sensitive to tactile stimuli with a fine hair are shown in Figure 3. Preparatory opening responses of tridentate and ophiocephalous organs can be elicited by tactile stimulation of the test surface or the outside of the valves, although these responses are less marked than the closing ones resulting from touching the inner surfaces of open valves. Reasonably consistent results were obtained for pedicellariae from *Prioncidaris baculosa, Echinus esculentus, Psammechinus miliaris, Tripneustes gratilla* and *Echinocardium cordatum.*

This information on the sensitivity of the valve epithelium for tridentate and ophiocephalous pedicellariae broadly agrees with the results of Von Uexküll's (1899) paper on echinoid pedicellarial physiology. Von Uexküll postulated a peripheral inhibitory chemical, present in the test epithelium, which inhibited tridentate and ophiocephalous organs from closing on other parts of the same animal. This he called 'autodermin'. Campbell & Laverack (1968) and Campbell (1974, 1976) failed to confirm the existence of an autodermin effect and found that under certain circumstances, e.g. when the inside surfaces

Table 3. Distribution of pedicellariae in echinoids

GLOBIFEROUS	Dactylous	'Claviform'	DENTATE	OPHIOCEPHALOUS	Claviform	PHYLLOUS = FOLIATE
Lacking in: Echinothuriids Salenids Aspidodiadematids Micropygids Usually lacking in: Clypeasteroids	Echinothuriids	In e.g. Diadematids Toxopneustids derived from globiferous organs by loss of head and retention of stem glands	Bidentate in Many clypeasteroids, especially in scutellids Tridentate in many regular echinoids Rostrate* spatangoids Quadridentate in saleniids Quinquedentates in laganids	Lacking in cidaroids Two types in *Arbacia* spp.	In e.g. *Salenocidaris* derived from ophiocephalous pedicellariae by loss of head and rention of stem glands	Lacking in cidaroids Triphyllous abundant in most other regulars except arbaciids Biphyllous in clypeasteroids ėspecially scutellids
Often absent in deep water regulars, those with venomous spines and burrowing irregulars	Restricted to deep water echinothuriids	Usually in warm water regulars, often those with globiferous organs with powerful venom	Widely distributed in echinoids of all habitats	Widely distributed in echinoids of all habitats	Usually in deep water echinoids	Widely distributed in echinoids of all habitats
Venom sac	Venom sac	Venom sac	No venom sec	No venom sac	Venom sac	No venom sac

* The status of rostrate pedicellariae is in need of review (M.Jensen, personal communication)

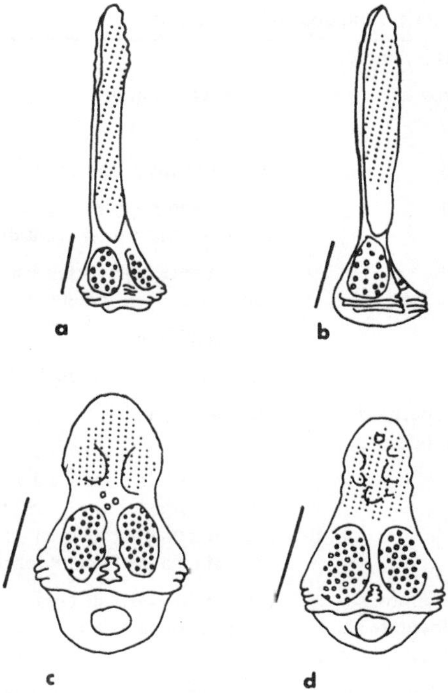

Figure 3. Areas of tridentate and ophiocephalous pedicellariae valves sensitive to tactile stimuli; a: *E.esculentus* tridentate, b: *P.miliaris* tridentate, c: *E.esculentus* ophiocephalous, d: *P.miliaris* ophiocephalous; dotted areas indicate tactile sensitivity on inside of jaws associated with closing responses; angled lines indicate vertical extent of sensitivity of outer face of valves associated with closing responses.

of valves were stimulated with a small spine, pedicellariae would attack another piece of the same animal.

Von Uexküll (1899) accurately noted that globiferous pedicellariae and the triphyllous pedicellariae were quite different from the other two echinoid classes. The former type, although sensitive to touch, close and release their venom only when an object bearing the appropriate chemical touches the sensory hillock. Tactile stimulation of the inside of these valves causes them to open further, thus exposing maximally the sensory hillock. This is a special aggregation of contact chemosensitive cells (Campbell & Laverack 1968) positioned relatively near the articulation point of the valves so that when struck, the intruding object itself will be so far in the grasp of the jaws as to be certain of being caught. There are different types of stimuli which affect globiferous pedicellariae (Fig.4), and it is logical that the autodermin effect would not be appropriate since touch on the inner jaw surface, even by part of the same individual's body, brings about a further opening rather than a closing response.

Triphyllous pedicellariae are different again. As Von Uexküll (1899) noted, these differ from all other echinoid types in their spontaneous activity. They spend most of their time browsing the test epithelium. As the rest of the test need not be protected against them, autodermin would not be appropriate. Their other principal difference is that they can move one valve separately

Table 4. Functions of pedicellariae.

(A) Echinoid		
Type	Proposed function	References
All	Parasites	Müller 1788
All	Young forms of the echinoid	L.Agassiz 1841
All	Seizing and gripping organs for passing food from one to another towards the mouth	Valentin 1841
All	Sort of antennae; to seize small animals for food, and to act in defence	Sars 1846
All	Organs of defence	Herapath 1865
All	Defence against fouling by fecal waste	A.Agassiz 1869
Unspecified (globiferous?)	Organs of defence	Perrier 1869
All	Removal of dirt and fecal waste; prehension; mucus from globiferous organs envelops small particles, making them more easy to dispose of: test cilia inadequate for particle removal	Sladen 1880
Globiferous	Involved in waste disposal	Romanes & Ewart 1881
Tridentate and Ophiocephalous	Assist in locomotion	
Tridentate and Ophiocephalous	Rapid jaw closure assists in grasping foreign bodies	Hamann 1887
Tridentate	Protect the test against larger organisms such as worms	
Globiferous	Grasping foreign bodies, their mucus secretion assisting in clutching weeds during locomotion and covering reactions	
Triphyllous	Cleaning	
All	Defence against fouling by settling organisms	Prouho 1887
Tridentate and Ophiocephalous	Capture of small organisms, e.g. settling larvae	Von Uexküll 1899
Globiferous	Defensive, using venom	
Triphyllous	Cleaning; removing small particles too small to be handled by the spines	
Globiferous	Defence against moderate and larger predators using venom	Prouho 1890, Fujiwara 1935, Jensen 1966
Tridentate	Picking up particles from test	Campbell 1973, 1974
Ophiocephalous	Catching moving objects	
Globiferous	Do not repel settling barnacle larvae; venom not involved	Campbell & Rainbow 1977
Triphyllous, Ophiocephalous and Tridentate	Effective with spines in discouraging settlement of barnacle cyprids	

Table 4 (continued).

(B) Asteroid		
All pedicellariae	Organs of prehension	Perrier 1869
*General: pedunculate pedicellariae	Dorsal pedicellariae of *Rathbunaster californicus* caught small crabs	Fisher 1906
Crossed pedunculate	Protection of papulae against the activities of small crustaceans	Jennings 1907
Straight pedunculate	Variable in response: capricious	
General: pedunculate pedicellariae	Capture of small fishes and crustaceans not necessarily used for food	Mauzey *et al.* 1968
Sessile	Did not capture commensal prawns unless contrived (in *A.planci*)	Campbell 1971
Crossed pedunculate	Prey catching	Robilland 1971 Chia & Amerongen 1975
*General pedicellariae	Respond to tactile and chemical extract (gastropod)	Lawrence 1982

* Did not specify which type

from the others. Still it is doubtful that they use it to grasp an object with two valves and smash it with the third as stated by MacBride (1906). Triphyllous pedicellariae do not seem especially sensitive to touch on the valves, although they can seize objects for brief periods. Touching the test seems to intensify their actions.

The histological evidence for the distribution of receptors generally fits in well with observations based on experimental work. Sladen (1880) was one of the first to identify a specific receptor area; the small but conspicuous sensory hillock on the inside of globiferous pedicellariae jaws. Hamann (1887) confirmed its existence. He observed the cilia, grouped closely together on the sensory hillock, and referred to them as 'sensory bristles' or 'tactile hairs'. Hamann also noted the existence of the distal receptor pad near the venom tooth in the globiferous pedicellariae. This feature has escaped the attention of some workers.

Following the activity of the late 19th century, a long period of inactivity occurred before new techniques enabled further progress on structure and function. Hyman's review of 1955 depended very largely on publications by 19th and early 20th century workers. Bargmann & Behrens (1963) were among the first workers to examine echinoderms using the transmission electron microscope and gave an account of the ultrastructure of the nerves of *Asterias rubens.* Cobb (1968b) investigated the globiferous pedicellariae of *E.esculen-*

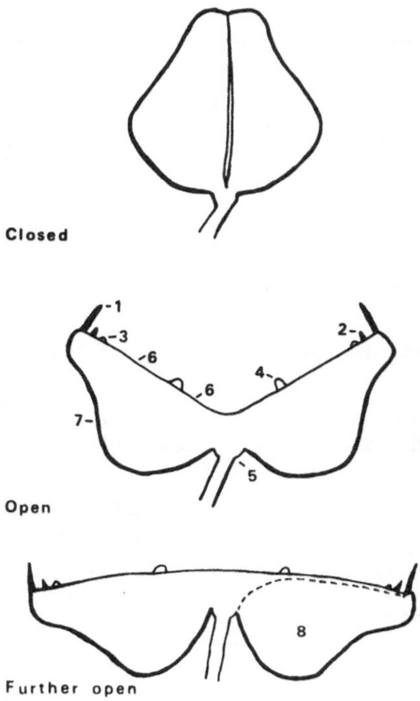

Figure 4. Areas of globiferous pedicellariae valves sensitive to tactile and contact chemical stimuli. These organs adopt three positions; closed, open and further open; 1: venom tooth; 2: accessory teeth; 3: distal tactile sensory pad; 4: contact chemosensory hillock, also sensitive to tactile stimuli; 5: point of articulation of head with top of stem; 6: inner face of valve sensitive to tactile stimuli; 7: outer face of pedicellaria represented by ephithelium of venom sac, this is weakly sensitive to touch; 8: extent of venom sac before venom discharge.

tus and found two types of epithelial cell, both of which he believed had a sensory role. Cobb effectively agreed with an earlier proposal of Kawaguti & Kamishama (1964) that all epithelial cells in echinoids are sensory. This proposition and the significance of cilia in sensory cells is beyond the scope of this review. Peters & Campbell (personal observation) have confirmed the distribution of these cell types in globiferous organs and their occurrence in other types of pedicellariae in *E.esculentus.* They also established that there is a distal sensory pad in globiferous pedicellariae and valve nerves in other types.

Oldfield (1975) found that cilia of the epithelium of globiferous pedicellariae from *P.miliaris* were liberally distributed on the inner surface of the valve, but that the outer surfaces bordering the venom sacs and the stem lacked cilia. Information on the distribution of receptors in other types of pedicellariae is needed.

3.2.2 *Asteroids*

Asteroid pedicellariae are poorly understood and little has been published regarding the distribution of their sensory cells. Jennings (1907) investigated the behavior of an asteroid and its pedicellariae in particular. In scope and style he seemed particularly influenced by the earlier works of Romanes (1884, 1885) and Von Uexküll (1899, 1900). His paper on *Asterias forreri (= Stylasterias forreri)* illustrates the difficulty of interpreting the perplexing behavior of asteroid appendages. He found that the crossed and straight pedicellariae of this species are sensitive to tactile stimuli by small crabs and other crustaceans and noted that isolated stimulation of the gills, papulae, or test had little effect on them. Repeated stimulation, however, brought large numbers of the organs into an open, alert position. Thereafter tactile stimulation of their inner jaw surfaces caused them to close and hold on to whatever they could catch. Fine sand and flucculent material had little effect upon their activity and most chemicals tried were too crude to produce meaningful results, although meat and juices from crabs excited them.

Campbell (1971) worked with the sessile pedicellariae of *Acanthaster planci* (Linnaeus) and with the straight and crossed pedicellariae of *Asterias rubens* (Linnaeus). The sessile pedicellariae of *A.planci* were sensitive to tactile stimuli. If touched with a bristle on the outside or on the test nearby, closed valves would open. The upper region of the inner valve epithelium were more sensitive to touch than the lower. A small spine taken from the asteroid produced similar results, confirming Jennings' observation that asteroids lacked the self-recognition, or autodermin, response that Von Uexküll reported for echinoids. The sessile pedicellariae appeared sensitive to movements of the small shrimp *Periclimenes sorror* Nobili which inhabits the test of *A.planci* in the Red Sea. These shrimps evaded the pedicellariae generally by sudden rapid movements but occasionally would be grasped by a pedicellaria and tear itself in the attempt to escape. In the case of *A.rubens,* results similar to Jennings' were obtained. The pedicellariae, both straight and crossed, opened in response to tactile stimulation of the nearby test or their outer surfaces, while their inner surfaces were also sensitive to touch. Campbell (1971) tested *A.rubens* with vibrational stimuli using sine waves at a range of frequencies from 0.75 Hz to 5.0 Hz but found no responses in the pedicellariae. Swimming nauplii larvae of *Artemia salina* (Linnaeus) were captured by the crossed pedicellariae but not by the straight pedicellariae. The activities of the caprellid amphipod *Pariambus typicus* (Krøyer), which lives in great numbers on the tests of some individuals of *A.rubens,* had little effect upon the state of the pedicellariae. These asteroids were capable of holding the caprellids if an artificial confrontation was brought about.

Our principal knowledge of the microanatomy of asteroid pedicellariae comes from the work of Smith (1937) who described the form of these organs in the context of a more general investigation of the nervous system. Asteroid pedicellariae have not been as well described as those of echinoids, and very little evidence is available on the precise distribution of receptors or the parameters they monitor. The opportunities afforded by ultrastructural analysis in this direction have yet to be fully exploited.

3.3 *Nervous system*

Pedicellariae are an attractive preparation for the behavioral neurobiologist because of the inbuilt reflex arc and the fact that in many echinoids they will function quite freely when detached from the test. Pentreath & Cobb (1972) gave a thorough review of the neurobiology of echinoderms and drew attention to the difficulties inherent in echinoderm tissues which have generally led to the failure of standard electrophysiological techniques. Such problems include small diameter axons and calcite crystals which make work with fine glass micro-electrodes very difficult. Several workers (including Von Uexküll 1896a, 1896b, 1900, Kinosita 1941, Bullock 1965, Smith 1965, 1966, Peters & Shelton 1981) have attempted to analyse the basi-epithelial nerve plexus which controls the test appendages of asteroids and echinoids. Once again, work with the echinoids has dominated, and it is the echinoid spine that has been the principal center of attention. Cobb (1970) produced a model of the echinoderm nervous system which he believed would apply to the peripheral system as well as to the radial cords. Campbell (1973) pointed out that previous efforts to analyse the basi-epithelial nervous system of the echinoid test had largely avoided pedicellariae and concentrated on spines, despite the fact that the former organs were much more numerous and certainly linked with the spines in response. Campbell developed the ideas of Jensen (1966) and applied his findings to those of Bullock (1965) in order to provide an explanation for control of pedicellariae. In essence, apart from the reflex responses, the general control of pedicellariae lies with the test system. Campbell pointed out that the ophiocephalous and tridentate pedicellariae were often held closed and lowered onto the test. They were alerted when tactile stimulation affected the test, their stems raised and their jaws opened. Further the ophiocephalous organs can converge on a center of tactile stimulation rather in the same way as do spines. Globiferous pedicellariae, on the other hand, are not involved unless the tactile stimulation is excessive so as to damage the epithelium. Extracts from the asteroid *A.rubens* selectively call the globiferous pedicellariae into alert behavior and bring about divergence of the spines and ophiocephalous pedicellariae which remain shut.

Peters & Shelton (1981) using neuropharmacological techniques, have shown that the spine-base neuromuscular system employs two different transmitters in excitation and inhibition.

It is the reflex activities of the valves of pedicellariae that have attracted more attention. Campbell & Laverack (1968) devised a method of recording the movements of isolated echinoid pedicellariae in order to monitor their responses to stimuli and determine some characters of the reflex arc. Campbell (1974, 1976) extended this work using *Echinus esculentus, Psammechinus miliaris, Prionocidaris baculosa, Tripneustes gratilla* and *Echinocardium cordatum.* Records were made using an adapted microscope with one eyepiece

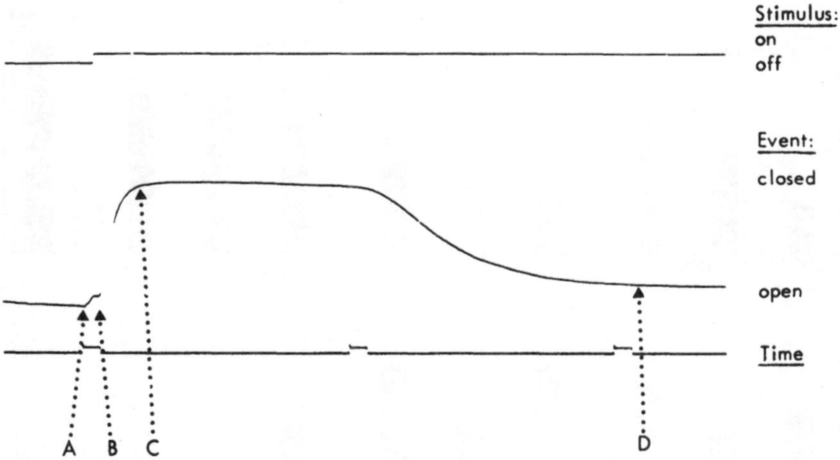

Figure 5. Foreign mechanical stimulation of an ophiocephalous pedicellaria of *Psammechinus miliaris* (specimen trace from the photoelectric apparatus). Latency is indicated by the arrowed points A-B; speed is indicated by the arrowed points B-C; response duration is indicated by the arrowed points C-D. The speed of this trace is 5 cm/sec.

replaced by a photoelectric cell. The pedicellariae were isolated from the test and mounted in the light path on the stage in a bath of sea water. A micromanipulator bearing a fine hair-probe was used to stimulate the jaws. The jaw movements resulting were monitored. As the jaws closed they progressively occluded the light path and thus produced a trace on the recorder (Fig.5). From such traces, it was possible to derive information on the latency between stimulation and response, the speed of the response, and its duration.

Figure 6 summarises the results for tridentate and ophiocephalous pedicellariae. It also shows that broadly similar results were obtained when the pedicellariae were stimulated with a fragment of another test appendage rather than with a foreign hair. A brief delay, characteristic of pedicellariae valves, occurs between the moment of stimulus and the initiation of a reflex closure response. This delay, or latency, increases if pedicellariae are exposed to serial stimulation at regular intervals, possibly due to antifacilitation in the reflex arc (Bullock & Horridge 1965; see Fig.7). This delay may reduce the efficiency of tridentate and ophiocephalous pedicellariae in trapping moving objects which repeatedly collide with them. In globiferous pedicellariae, Campbell (1976) found that serial tactile stimuli applied to the inner surfaces of open jaws produced a further opening response where the latency did not increase. Perhaps these pedicellariae are adapted to make numerous preparatory responses prior to receiving contact chemostimulation which will induce closure and venom release. The closure with venom-release response for globiferous pedicellariae is not repeatable and the effect of serial repetition on its latency could not be

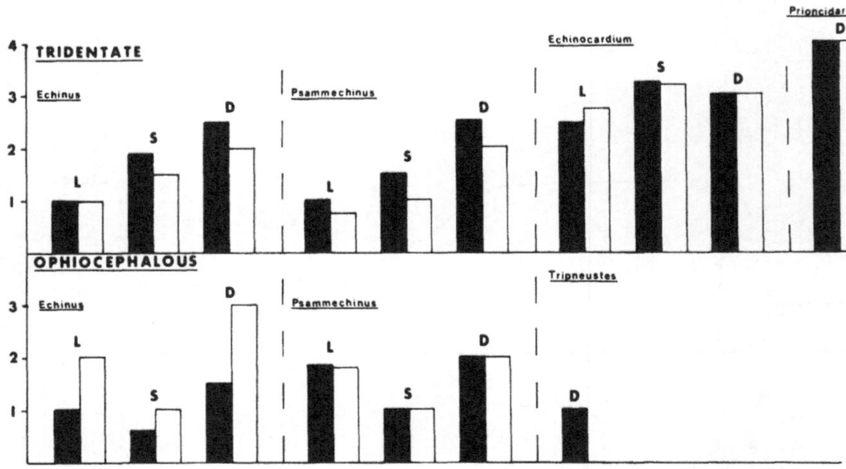

Figure 6. Histogram summarising the data for jaw responses to tactile stimulation by a foreign object (solid boxes) and another part of the echinoid's own body (open boxes) for 5 species as marked: *Prionocidaris baculosa, Echinus esculentus, Psammechinus miliaris, Tripneustes gratilla* and *Echinocardium cordatum.*

tested. Triphyllous pedicellariae were too small to yield results in the recording apparatus, so a comparison of their valve movements cannot as yet be made.

Hamann's (1887) contribution to the microanatomy of echinoid pedicellariae demonstrated quite clearly the existence of nerves running from the test plexus up the stem. It also showed a well-defined nerve running from areas of receptor cells to the muscles of ophiocephalous pedicellariae. This nerve has been referred to by Campbell (1972) as the valve nerve. He was not able to relate fully the innervation of the adductors and the abductors with the nerves of the stem. Peters & Campbell (personal observation) have investigated the form of the nervous system in the heads and stems of all four pedicellariae from *E.esculentus* (Fig.8). It is clear from the literature that a detailed knowledge of the form of the skeletal elements is essential before the insertion of the muscles can be finally established. Only when this has been done can the precise course of the valve nerve and its branches be charted. The various histological studies of Cobb (1968a,b), Cannone (1970) and Chia (1969a) have proved very helpful in establishing the layout of the head and the positions of the nerves. Each valve in echinoids (Fig.8) has an area of neuropil (defined by Bullock & Horridge 1965, as 'a tangle of fine fibres and their endings, quite or nearly devoid of nerve cell bodies, . . . and forming much of the bulk of invertebrate ganglia, where it is usually the sole synaptic field . . .'). Such areas of neuropil play a major role in the integrative processes of echinoderm nervous systems (Cobb 1970, Pentreath & Cobb 1972). This neuropil, enclosed

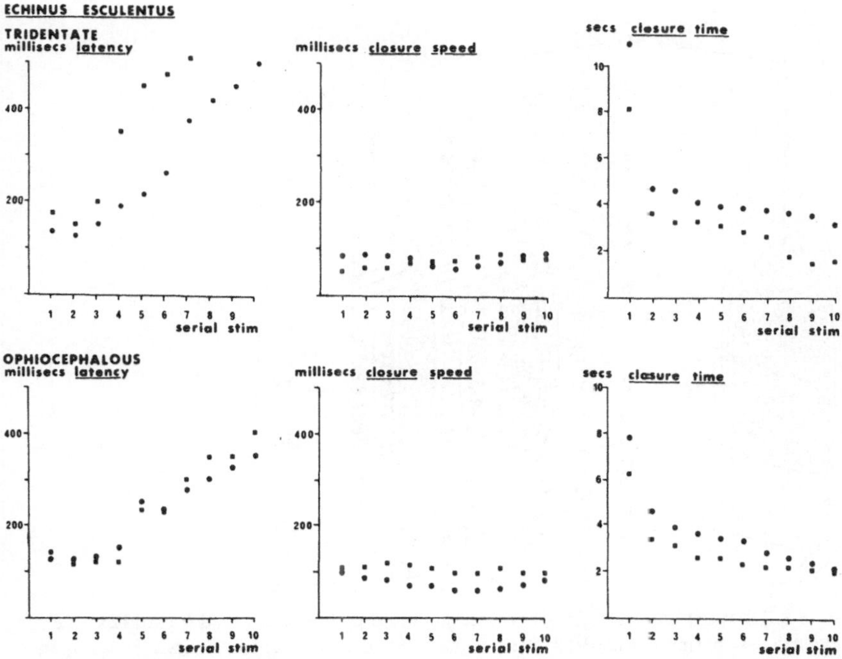

Figure 7. The effect of serial stimuli repeated at regular intervals on pedicellariae from *E. esculentus.* Circle: foreign tactile stimulus; square: tactile stimulus by another part of the echinoid's body.

inside the hollow cavity of the keel of each valve, is well protected by the skeleton. Cobb (1968a) showed that the adductor muscles of tridentate pedicellariae send processes through the meshwork of the inner face of the ossicle at this level to receive their innervation, and he demonstrated synaptic and other vesicles likely to play a part in the innervation of the muscles. This neuropil appears responsible for integrating the activities of the two sets of synergistic adductor muscle fibers, abductor muscles and flexor muscles. There is but one set of adductors in the globiferous type, but the venom discharge musculature has to be controlled. There are striated and smooth adductor muscles in the triphyllous pedicellariae, which require integration. The spontaneous activity in the valves, characteristic of this type, requires that a pacemaker, or some impulse generating mechanism, must reside in the pedicellaria head. Spontaneous activity can persist after detachment of the pedicellariae from the test, so it is unlikely that the pace-maker function lies in the basiepithelial plexus.

Echinoderm nervous systems are renowned for their apparent lack of ganglia. Classical histologists like Baudelot (1872) and Prouho (1887) noted the prob-

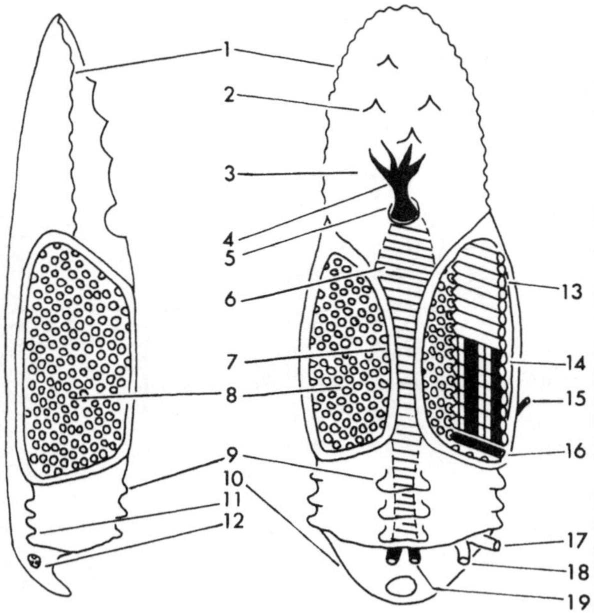

Figure 8. Essential features of a generalised echinoid pedicellaria based on studies on *E. esculentus.* 1: peripheral teeth; 2: central teeth; 3: receptors; 4: valve nerve; 5: foramen; 6: area of neuropil; 7: keel; 8: adductor muscle insertion region; 9: central cog-teeth; 10: handle; 11: lateral cog-teeth; 12: abductor and flexor insertion areas; 13: smooth adductor muscle fibers; 14: striated adductor muscle fibres; 15: venom sac nerve (globiferous pedicellariae only); 16: inter-valve link nerve; 17: smooth abductor; 18: smooth flexor; 19: nerves to stem.

lem of definite identification of ganglia in the form of aggregations of cell bodies, although Prouho did succeed in showing cell bodies in the basi-epithelial plexus and spine bases of *Echinus acutus.* Smith (1937), working on asteroids, described 'ganglion cells' in the radial nerve cords of *Marthasterias glacialis.* There have been ultrastructural descriptions of multipolar nerve cell bodies (Pentreath & Cottrell 1971, Weber & Grosmann 1977, Bachmann & Goldschmid 1978b). According to Bullock & Horridge (1965) a ganglion is 'a discrete collection of nerve cells' often 'a nodular mass defined by connective tissue'. It would seem that Cobb & Laverack's (1966) account of the hyponeural tissue of *E.esculentus* satisfies this definition. Chia (1970) described a ganglion in the globiferous pedicellaria of *P.miliaris.* This comprised about 75 aggregated cells in the distal portion of the valve nerve. Three similar structures have been located at intervals in each valve of the globiferous pedicellariae of *E.esculentus;* immediately below the distal sensory pad, midway between the pad and the hillock, and at the top of the keel (Peters & Campbell, unpublished observation). Whether or not these aggregations of nerve cell bodies are true ganglia is still open to debate. What of the ultrastructure of the axons themselves? Many descriptions of echinoderm nerves have been reviewed by Pentreath & Cobb (1972). Subsequent contributions have been made by Brehm (1977) and Weber & Grosmann (1977). The axons of peripheral systems in echinoids resemble those of the radial nerve cords, except that they are not arranged in tightly packed tracts (Cobb 1968b, Weber & Grosmann 1977). Generally the axons are small and unspecialised. Peters (1981) reported no clear distinction in the basi-epithelial plexus between sensory and motor axons on the basis of size, despite the claims of Smith (1965, 1966) who believed that the plexus arrangements in echinoids would be similar to those known for asteroids and holothuroids. The identification of motor neurons in pedicellariae has been rather difficult on account of the closeness of the innervation of the muscles with the neuropile of the keel. However one site suitable for this investigation is the nerve of the venom sac muscle which has not been scrutinised ultrastructurally. It will be noted (Fig.8) that the neuropil masses of each keel are interconnected by the valve link nerve whose role is presumably to spread excitation of impulses generated by one receptor to the two other areas of neuropile in the head. The arrangement of the nervous system of pedicellaria bases has received little attention hitherto despite the clear importance of pedicellaria stems in most responses in echinoids. Peters (1981) found that, in most respects, they resemble those of spines on a smaller scale. This supports Hyman's belief (1955) of the morphological equivalence of the two types of appendage.

There is great need for research into the ultrastructure of spine and pedicellaria bases in the irregular echinoids, with their wealth of spine modifications. Smith (1937) using light microscopy has provided the only account of the nervous elements in asteroid pedicellariae. Other workers, reviewed by Pentreath & Cobb (1972) and Cobb & Raymond (1979), have given ultrastructural accounts of asteroid nerves.

3.4 *Muscle systems*

The existence of muscles to open and close the jaws of pedicellariae is one of the most apparent features of these organs. The suggestion that the valves were moved by antagonistic sets of adductors and abductors was first made by Herapath (1865). Geddes & Beddard (1883) and Beddard (1886) described the arrangement of adductor and flexor muscles of echinoids, although Sladen (1880) had already discovered the former. Koehler (1883) confirmed the existence of the adductors but did not notice, as had Geddes & Beddard (1883), that both tridentate and ophiocephalous pedicellariae contained a striated element in the adductor block. Hamann (1887), who figured the abductors and provided a satisfactory explanation for the role of the flexors in moving the whole head in relation to the stem, claimed incorrectly that he was the first to discover striated elements. Recent work has shown that triphyllous pedicellariae contain striated muscles (Peters 1981). Striated muscle in pedicellariae has been misinterpreted. For example, Campbell & Laverack (1968) figured a striated filament lying beside a smooth one in tridentate pedicellariae. The striated nature of the 'smooth' muscle was obscured by the plane of sectioning. Smooth muscles vary little from place to place in echinoid pedicellariae. They are a more constant feature and generally have larger fibrils than the striated muscles (Peters & Campbell unpublished observation). Many papers describe the venom sacs of globiferous pedicellariae (Hamann 1887, Lewis & Saluja 1967, Cannone 1970, Chia 1970, O'Connel *et al.* 1974, Holland & Holland 1975) but give little information on the arrangement of the venom sac muscles themselves. The disposition of the muscle types is given in Figure 8. We are as ignorant of the ultrastructure of asteroid pedicellarial muscles, as we are concerning details of their activities.

The experiments of Campbell & Laverack (1968) and Campbell (1974, 1976) provide information on the speed of valve movements in the two of the three larger pedicellariae from *E.esculentus* as well as on the duration of closure (Fig.7). Campbell & Laverack (1968) and Campbell (1974, 1976) investigated the function of adductor and abductor muscles in tridentate, ophiocephalous and globiferous pedicellariae. They measured closure speeds and closure durations with the photoelectric recording apparatus referred to above. Closure speeds initiated by both foreign and self-derived tactile stimuli are rapid and reasonably constant (Fig.7). The closure speed, at around 100 milliseconds, would seem adequate to snap the jaws shut before the stimulus source had moved out of range. Serial stimulation shortens closure duration. This suggests a dual system in which the striated adductor element takes responsibility for the rapid closure movement, and the smooth element for holding the jaws shut. The results of micro-surgery supported this view.

It is probable that these pedicellariae are not evolved to make repeated closures and that the system which holds the valves shut fatigues quite rapidly, hence the decline in closure time to repeated stimuli (Fig.7).

The further opening response of globiferous pedicellariae does not show a deterioration with repeated stimulus: probably this is an adaptation to remaining 'at the ready' in anticipation of a contact-chemostimulus leading to a closure and venom release. The discovery that triphyllous pedicellariae from *E.esculentus* have striated elements makes this doubtful; perhaps these pedicellariae function in some way other than by seizing.

The muscle systems of asteroid pedicellariae are not well known. Smith (1937) and Chia & Amerongen (1975) have provided a detailed account of the muscle layout. There appear to be two sets of adductors and two sets of abductors in *Stylasterias forreri*, providing a most sophisticated mechanism for the movement of the valves. Apparently there is some difference in the layout of the adductor muscles of *Asterias rubens* and *Stylasterias forreri*. The extent of such differences in forcipulate crossed pedicellariae should be investigated further. Little information is available on the functioning of the muscles themselves. Campbell (1971) found that sessile pedicellariae from *Acanthaster planci* had an average closure duration to tactile stimulus of 4.5 seconds when the stimulus was not captured. If the prey was captured, the average closure duration was 10 seconds. For straight pedicellariae from *Asterias rubens,* the average closure duration when the probe was retained was 100 seconds. This is much longer than for the crossed pedicellariae, where the average closure duration was 3.5 seconds. Although the closure speed could not be measured, these figures suggest a differential capacity and therefore possibly a different function in these two types.

3.5 *The skeleton*

Campbell (1972) used the scanning electron microscope to investigate the valves of pedicellariae from *Echinus esculentus* and further work on other species was carried out by Jensen (1974), Oldfield (1976), Hilgers & Splechtna (1976, 1979 and 1981) and Splechtna & Hilgers (1980). In asteroids this technique was used by Campbell (1971) for *Acanthaster planci* and *Asterias rubens.* Chia & Amerongen (1975) looked at *Stylasterias forreri* (see Fig.9).

The finely constructed articulation points of the valves ensure precise meeting at their tips (in tridentate valves) and almost all round the periphery (in ophiocephalous and triphyllous valves), thus rendering these pedicellariae capable of seizing efficiently. Without such an arrangement pedicellariae would be useless at handling small objects. Disarticulation in tridentate and ophiocephalous valves cannot occur without breakage of the pivot system. The jaws disarticulate in the further opening responses of globiferous pedicellariae and during the independent jaw movements of triphyllous pedicellariae. Campbell (1972) referred to the articulation points as cog-teeth because of their resemblance to arcs of meshing cog wheels.

The skeleton is also highly modified for the insertion of the adductor, abductor and flexor muscles as well as for the protection of the valve nerve and neuropile inside the keel. The main features of the echinoid valve skeleton are summarised in Figure 8. Collagen fibers link the stem osicle with the valves. These fibers in the tridentate and ophiocephalous pedicellariae (Campbell 1972, Splechtna & Hilgers 1980) pass over a part of the ossicle referred to as the handle. This transmits any increase in tension in the flexible part of the stem, such as may be carried by a struggling captive, to the valves, so making them close harder. By this means the gripping power of the pedicellaria is not limited by the power of the adductors. A struggling captive, thrusting against the test, will generate tension in the stem and thus imprison itself more firmly.

A remarkable analogy is found between this arrangement and that of the crossed pedicellariae of asteroids (Campbell 1971, Chia & Amergongen 1975) where the arrangement of the ossicles and connective tissue fibrils achieves a similar effect. Descriptions of skeletal details of valves of pedicellariae of many asteroids and echinoids were made by Perrier (1869) and Mortensen (1928-1951, 1932), and thus skeletal details are available for virtually every species recognised.

3.6 *Nature of pedicellarial venom*

The pioneer work on toxicology of pedicellariae was carried out by Von Uexküll (1899) and Henri & Kayaloff (1906). They reported its paralysing and fatal effects on various marine invertebrates including gastropods and crustaceans. Since their publications appeared pedicellarial venom has received little attention. Fujiwara (1935) described the poisonous effects of the globiferous pedicellariae from *Toxopneustes pileolus.* He noted the paralysing effects and severe pain he experienced when bitten and recorded respiratory difficulties and a fall in body temperature in mice when they were attacked. Pérès (1950) found the active fraction of venom from *Sphaerechinus granularis* to be thermostable but Mendes *et al.* (1963) found it to be heat labile and dialyzable. Alender (1964), using *Tripneustes gratilla* noted the responses to globiferous pedicellariae and found active factors to be heat labile but non-dialyzable. Using the same species, Feigen *et al.* (1966) found such factors to contain proteins which acted as enzymes in immunochemical experiments. O'Connell *et al.* (1974) investigated the histology of the venom gland cells from globiferous pedicellariae of *Strongylocentrotus purpuratus.* The venom gland cells appeared mucoid and their cytoplasm contained conspicuous vesicles and numerous cytomembrane complexes indicating metabolic activity immediately prior to fixation. The products of the vesicles appeared to consist of protein and non-sulfated acid muco-substances. These authors believed that the gland cells release their vesicle complement when the venom sac muscles contract.

Less than 0.2 % of references cited by Habermehl & Mebs (1981) in the extensive bibliography of poisons, toxins and venoms of natural origin relate to echinoderms. This underlines the need for more research in this field, especially if we are to understand the selective advantages of highly venomous pedicellariae in echinoids.

3.7 *Development and regeneration*

Pedicellariae from adult echinoderms have attracted the attention of researchers, but these organs exist on larvae of both asteroids (Downey, personal communication) and echinoids (Burke, personal communication).

Although Mortensen (1938) described the development of echinoderm larvae, much remains to be discovered on the role of pedicellariae in larval and juvenile forms. Baker & Campbell (personal observation) found that all classes of pedicellariae do not appear simultaneously in the larvae and juveniles of *E. esculentus*.

Pedicellarial regeneration has been observed in a number of instances. Pérès (1950) reported that the gemmiform pedicellariae of *Sphaerechinus granularis* had great powers of regeneration, and Chia (1970) provided a study of the regeneration of globiferous pedicellariae in *Psammechinus miliaris.* Since globiferous pedicellariae autotomise after the discharge of venom, it is therefore necessary for the animal to replace them as rapidly as possible in order to maintain an effective defence. Chia (1970) found that a period of about 25 days was required for a new globiferous pedicellarial head to be regenerated, at which point it had reached its maximum size and was fully functional; however, a further period of about 15 days was required during which time the process of stalk elongation was completed.

Chia stated that pedicellarial regeneration is not synchronised, and he had not at this point established whether or not pedicellariae regenerated on their sites. Regeneration is retarded several days if all external appendages are removed, since wound healing must first occur, and the appendages of the aboral half grow faster than those of the oral half.

4 ROLE OF PEDICELLARIAE: ENVIRONMENTAL ASPECTS

Because of their small size, isolated pedicellariae do not lend themselves to field investigations. Nevertheless, Campbell & Rainbow (1977) concluded that cyprid larvae of the acorn barnacle *Semibalanus balanoides* were prevented from settling on live *Echinus esculentus.* There was no chemical repulsion of cyprids by any substance of the test espithelium and the animal's surface itself was suitable for cyprid settlement. Thus the protection against fouling by cyprids and similar larvae lie in the activities of the spines and/or the pedicellariae. The aboral region of the echinoid is most favored for settlement although in the experiments the oral/aboral axes were not placed vertically. The pore surfaces of denuded tests were not as attractive, probably because of their size.

Ramsay & Campbell (unpublished observations) found that the populations of pedicellariae do not vary very much longitudinally around the test but they do vary latitudinally. Triphyllous pedicellariae appear evenly distributed except that there are fewer in the interambulacrum which bears the madreporite. Ophiocephalous pedicellariae are evenly distributed all round the animal but the tridentate pedicellariae show a variation with height. Aborally these are significantly more numerous in the two ambulacra flanking the madreporite; at the ambitus they are evenly distributed all round, and orally there is a confusing discrepancy between the ambulacra and the interambulacra. Finally, the globiferous pedicellariae, which one would expect to be variable because of their loss after use, are significantly more numerous in the interambulacrum which bears the madreporite. Perhaps pedicellarial populations on an individual urchin may relate to its orientation either towards water currents or in relation to the direction of travel. More information on the real role of the madreporite would be of use in this respect.

The discovery on *E.esculentus* of large thigmotrich ciliates (Campbell & Peters, personal observation) attached to the inner surface of the valves of triphyllous pedicellariae recalls the protozoans and other micro-organisms found in the echinoid gut (Ferguson 1969). Possibly these organisms are feeding on bacteria scraped up from the exterior surface of the urchin by the triphyllous pedicellariae, and that bacterial collection may be a role of this type of appendage.

Finally mention must be made of the question of 'skin digestion' and the nutrition of pedicellariae. The valves and active musculature are borne on stems in many cases and there are no obvious links with distributive systems of the animal body. Either the products of gut digestion must be carried to the pedicellariae by wandering cells, of which there are no shortages, or the pedicellariae must obtain nutrients directly from seawater. Ferguson (1970) found that the epithelium of the test of *Echinaster echinophorus* received very little, if any benefit, from the injection of nutrient ^{14}C labelled amino acid into the central region of the body cavity and the stomach lumen. Péquignat (1970) suggested that the pedicellariae of echinoids might be fed either directly from the environment, or as a result of digestion of small food organisms trapped against the surface of the epithelial cells in the presence of enzymes (Péquignat 1966). Stephens (1968) suggested that food substances within seawater can support the metabolic demands of organisms by direct absorption. It is in the light of such facts that the ultrastructure of the asteroid and echinoid test epithelium, with its microvilli resembling gut epithelium (Cobb 1968b) takes on a greater significance. Pedicellariae are covered with such epithelium (Campbell 1971).

5 CONCLUSION

Most research has concentrated on the pedicellariae of regular echinoids. The gross anatomy and fine structure of these organs is well documented, but there are many important questions relating to their function and metabolic relationships, especially to their defence potential, grooming ability and connection with feeding. Field studies could contribute much, as could development and regeneration studies.

Knowledge of irregular echinoids is much less complete. There are only brief accounts of the structure and function of their pedicellariae, in many cases based on similarities with regular echinoids. Because their life-styles are so different, some assumptions of pedicellarial roles in these echinoids may be false.

There are few up-to-date accounts of the structure and function of asteroid

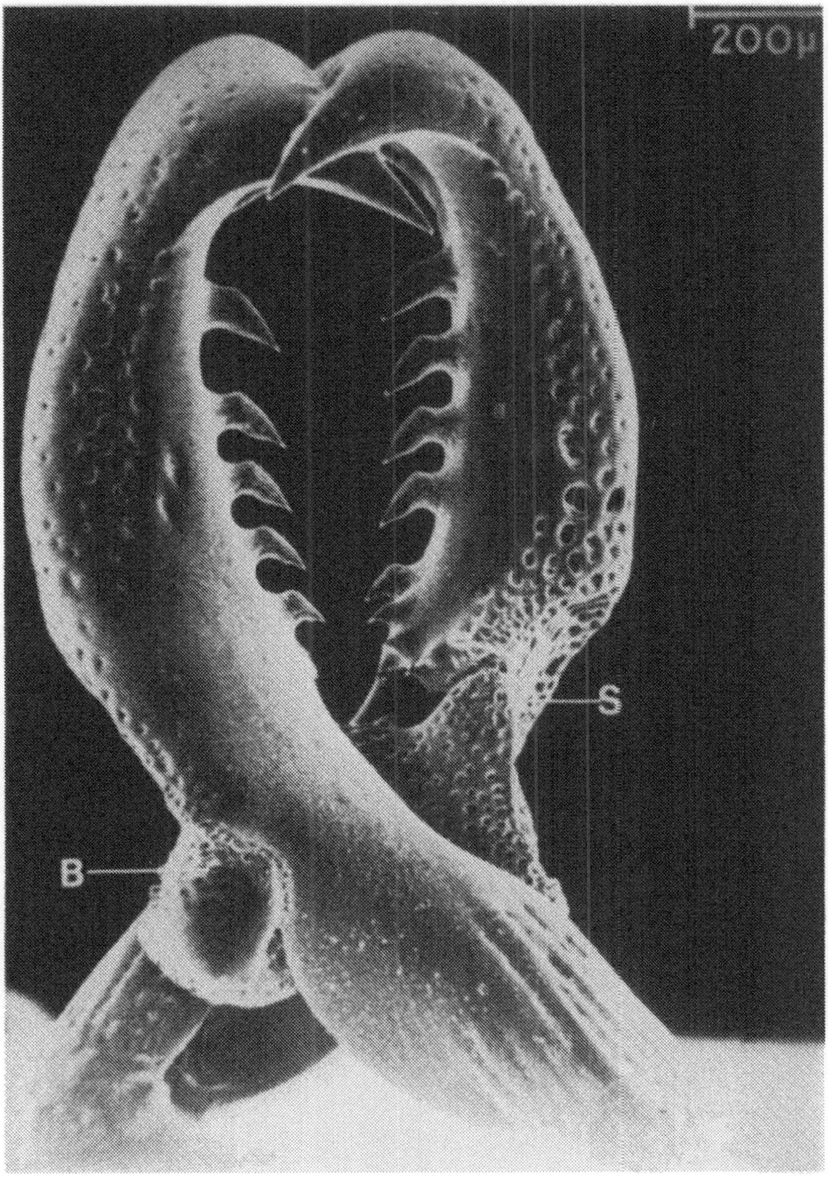

Figure 9. Scanning electron micrograph of the skeletal parts of the crossed pedunculate pedicellaria from *Stylasterias forreri* (reproduced with kind permission from Chia & Amerongen 1975).

pedicellariae. There is a great need for research into the micro-anatomy, ultra-structure and neurobiology of asteroid pedicellariae.

6 ACKNOWLEDGEMENTS

I am particularly grateful to Dr B.H.Peters and to Miss E.K.Ramsay who have allowed me to quote from their as yet unpublished results. Thanks are also due to Miss M.E.Downey, Miss C.Ellis, Dr M.Jensen and Professor D.Nichols, for their helpful criticism of the manuscript.

7 REFERENCES

Agassiz, A. 1869. Preliminary report on the echini and starfishes dredged in deep water between Cuba and the Florida reef. *Bull. Mus. comp. Zool. Harv. I.* 9.

Agassiz, L. 1841. *Monographies d'Echinodermes vivans et fossiles.* Neuchâtel.

Alender, C.B. 1964. The venom from the heads of globiferous pedicellariae of the sea-urchin *Tripneustes gratilla* (Linnaeus). Ph.D. thesis, University Hawaii. 140pp.

Bachmann, S. & A.Goldschmid 1978. Ultrastructural fluorescence microscopical and microfluorimetric study of the innervation of the axial complex in the sea urchin *Sphaerechinus granularis* (Lam.). *Cell. Tiss. Res.* 194:315-326.

Bargmann, W. & B.Behrens 1963. Uber den Feinbau des Nervensystems des Seesternes (*Asterias rubens* L.) II Mitteilung. Zur Frage des Baues und der Innervation der Ampullen. *Z. Zellf.* 59:746-770.

Baudelot, E. 1872. Etudes générales sur le système nerveux. *Archs Zool. exp. gen.* 1:177-216.

Beddard, F.E. 1886. Striated muscles in Echinida. *Ann. Mag. nat. Hist.* 17:428-430.

Brehm, P. 1977. Electrophysiology and luminescence of an ophiuroid radial nerve. *J. exp. Biol.* 71:213-227.

Bullock, T.H. 1965. Comparative aspects of superficial conduction systems in echinoids and asteroids. *Am. Zool.* 5:545-562.

Bullock, T.H. & G.A.Horridge 1965. *Structure and function in the Nervous Systems of Invertebrates.* W.H.Freeman & Co, San Francisco.

Campbell, A.C. 1971. Studies on some effector organs in echinoderms. Ph.D. Thesis, University Oxford.

Campbell, A.C. 1972. The form and function of the skeleton in pedicellariae from *Echinus esculentus* L. *Tiss. Cell,* 4:647-661.

Campbell, A.C. 1973. Observations on the activity of echinoid pedicellariae: I. Stem responses and their significance. *Mar. Behav. Physiol.* 2:33-61.

Campbell, A.C. 1974. Observations on the activity of echinoid pedicellariae: II. Jaw responses of tridentate and ophiocephalous pedicellariae. *Mar. Behav. Physiol.* 3:17-34.

Campbell, A.C. 1976. Observations on the activity of echinoid pedicellariae: III. Jaw responses of globiferous pedicellariae and their significance. *Mar. Behav. Physiol.* 4:25-39.

Campbell, A.C. & M.S.Laverarck 1968. The responses of pedicellariae from *Echinus esculentus* (L.). *J. exp. mar. Biol. Ecol.* 2:191-214.

Campbell, A.C. & P.S.Rainbow 1977. The role of pedicellariae in preventing barnacle settlement on the sea-urchin test. *Mar. Behav. Physiol.* 4:253-260.

Cannone, A.J. 1970. The anatomy and venom-emitting mechanism of the globiferous pedicellariae of the urchin *Parechinus angulosus* (Leske) with notes on their behaviour. *Zool. Afric.* 5:179-190.

Chia, F.S. 1969a. Histology of the pedicellariae of the sand dollar *Dendraster excentricus* (Echinodermata). *J. Zool., Lond.* 157:503-507.

Chia, F.S. 1969b. Responses of the globiferous pedicellariae to inorganic salts in three regular echinoids. *Ophelia* 6:203-210.

Chia, F.S. 1970. Histology of the globiferous pedicellariae of *Psammechinus miliaris. J. Zool. Lond.* 160:9-16.

Chia, F.S. & H.Amerongen 1975. On the prey-catching pedicellariae of a starfish, *Stylasterias forreri* (de Loriol). *Can. J. Zool.* 53:748-755.

Cobb, J.L.S. 1968a. The fine structure of the pedicellariae of *Echinus esculentus* (L.). I. The innervation of the muscles. *J. Roy. micr. Soc.* 88:211-221.

Cobb, J.L.S. 1968b. The fine structure of the pedicellariae of *Echinus esculentus* (L.). II. The sensory system. *J. Roy. micr. Soc.* 88:223-233.

Cobb, J.L.S. 1970. The significance of the radial nerve cords in asteroids and echinoids. *Z. Zellf.* 108:457-474.

Cobb, J.L.S. & M.S.Laverack 1966. The lantern of *Echinus esculentus* (L.). II. The fine structure of the hyponeural tissue and its connections. *Proc. Roy. Soc. Lond.* (B)164: 651-658.

Cobb, J.L.S. & A.M.Raymond 1979. The basi-epithelial plexus of the viscera and coelom of eleutherozoan Echinodermata. *Cell. Tiss. Res.* 202:155-163.

Cuénot, L. 1887. Contribution a l'étude anatomique des astérides. *Arch Zool. exp. gén.* (2)5 bis:1-144.

Doederlein, L. 1887. *Die japanischen Seeigeln I. Die Familien Cidaridae und Saleniidae.* Stuttgartt. 59pp.

Doederlein, L. 1906. Die Echinoiden der Deutschen Tiefsee-Expedition. *Wiss. Ergebn. Dtsch. Tiefsee-Exped. Valdivia,* 5(2).

Duvernoy, M. 1849. Mémoire sur l'analogie de composition et sur quelques points de l'organisation des échinodermes. *Mém. Acad. Sci. Paris* 20:64pp.

Feigen, G.A., E.Sanz, J.T.Tomita & C.B.Alender 1966. Studies on the mode of action of sea-urchin toxin. II. Enzymatic and immunological behavior. *Toxicon* 6:17-43.

Ferguson, J.C. 1969. Feeding, digestion and nutrition in Echinodermata. In: M.Florkin & B.T.Scheer (eds), *Chemical Zoology* 3:71-100. Academic Press, New York.

Ferguson, J.C. 1970. An autoradiographic study of the translocation and utilisation of amino acids by starfish. *Biol. Bull.* 138:14-25.

Fisher, W.K. 1906. The starfishes of the Hawaiian Islands. *Bull. U.S. Fish Comm.* 1903: 987-1130.

Fisher, W.K. 1928. Asteroidea of the North Pacific and adjacent waters. *Bull. U.S. natn. Mus.* 76:1-161.

Foettinger, A. 1881. Sur la structure et la forme des pédicellaires gemmiformes de *Sphaerechinus granularis. Archs Biol. Paris.* 2:450-492.

Fujiwara, T. 1935. On the poisonous pedicellariae of *Toxopneustes pileolus. Annotn. zool. Jap.* 15:62-69.

Geddes, P. & F.E.Beddard 1883. On the histology of the pedicellariae and the muscles of *Echinus sphaera* (Forbes). *Trans. Roy. Soc. Edinburgh* 30:383-396.

Gosse, P.H. 1856. *Tenby, a Sea-side Holiday.* J.van Voorst, London. 400pp.

Habermehl, G. & D.Mebs 1981. 1980 Bibliographic index: poisons, toxins and venoms of natural origin. *Toxicon* 19:583-603.

Hamann, O. 1887. *Beitrage zur Histologie der Echinodermen Heft 3 Anatomie und Histologie der Echiniden und Spatangiden,* Gustav Fischer, Jena.

Haubold, E. 1933. Uber eine neue Form sitzender Pedizellarien bei Seesternen. *Zool. Anz.* 103:199-205.

Henri, V. & E.Kayaloff 1906. Etude des toxines contenues dans les pédicellaires chez les oursins. *C.R. Soc. Biol. Paris* 60:884-886.

Herapath, W.B. 1865. On the pedicellariae of Echinodermata. *Quart. J. microsc. Sci.* 5: 175-185.

Hilgers, H. & H.Splechtna 1976. Struktur – und Funktion – analyze ophiocephaler Pedizellarien von *Sphaerechinus granularis* Lamarck, *Echinus acutus* Lamarck und *Paracentrotus lividus* Lamarck (Echinodermata, Echinoidea). *Zoomorph.* 86:61-80.

Hilgers, H. & H.Splechtna 1979. Die Entwicklung ophiocephaler Pedizellarien von *Sphaerechinus granularis* (Echinodermata, Echinoidea). Strukturelle und funktionelle Analyse. *Zoomorph.* 93:265-287.

Hilgers, H. & H.Splechtna 1981. Zur Feinstruktur ophiocephaler Pedizellarien von *Arbacia lixula* (Linne) (Echinodermata, Echinoidea). *Zoomorph.* 97:89-100.

Holland, L.Z. & N.D.Holland 1975. The fine structure of epidermal glands of regenerating and mature globiferous pedicellariae of a sea urchin *(Lytechinus pictus)*. *Tiss. Cell.* 7:723-737.

Hyman, L.H. 1955. *The Invertebrates. Vol IV. Echinodermata.* McGraw-Hill, New York.

Jennings, H.S. 1907. Behaviour of the starfish *Asterias forreri* de Loriol. *Univ. Calif. Publ. Zool.* 4:53-185.

Jensen, M. 1966. The responses of two sea-urchins to the sea star *Marthasterias glacialis* (L) and other stimuli. *Ophelia* 3:209-220.

Jensen, M. 1974. The Strongylocentrotidae (Echinoidea). A morphologic and systematic study. *Sarsia* 57:113-148.

Kawaguti, S. & Y.Kamishama 1964. Electron microscopic study of the integument of the echinoid *Diadema setosum. Annotnes Zool. Jap.* 37:147-152.

Kiernik, E.von 1905. Beitrag zur Histologie der Pedicellarien der Echiniden, insbesonders der Muskeln. *Zool. Anz.* 29:610-614.

Kinosita, H.J. 1941. Conduction of impulse in superficial nervous system of sea urchin. *Jap. J. Zool.* 9:221-232.

Koëhler, R. 1883. Recherches sur les échinides des côtes de Provence. *Annls Mus. Hist. nat. Marseille* 1(3):1-167.

Lawrence, J.M. 1982. Stimulation of the pedicellariae of *Coscinasterias tenuispina* (Lamarck) (Echinodermata: Asteroidea). *Fla. Sci.* 45(suppl.1):20.

Lewis, J.B. & G.Saluja 1967. The claviform pedicellariae and their stalk glands in the tropical sea urchin *Diadema antillarum* Philippi. *Can. J. Zool.* 45:1211-1214.

MacBride, E.W. 1906. Echinodermata. In: *Cambridge Natural History,* vol.1. Cambridge University Press.

Mauzey, K.P., C.Birkeland & P.K.Dayton 1968. Feeding behaviour of asteroids and escape responses of their prey in the Puget Sound region. *Ecology* 49:603-619.

Mendes, E.G., L.Abbud & S.Vmiji 1963. Cholinergic action of homogenates of sea-urchin pedicellariae. *Science* 139:408-409.

Mortensen, T. 1903. Echinoidea Pt.1. *Danish Ingolf Expedition* 4:1-193.

Mortensen, T. 1928-1951. *Monograph of the Echinoidea* 15 vols. C.A.Reitzel, Copenhagen.

Mortensen, T. 1932. Ueber den angeblichen Kieselschwamm *Microcordyla asterial* Zinpolo. *Zool. Anz.* 97:197-204.

Mortensen, T. 1938. Contributions to the study of the development and larval forms of echinoderms IV *Kong. Danske Vidensk. Selsk. Skrift., Naturv. Math Afd.* (9)7(3):1-59.

Müller, O.F. 1788. *Zoologia Danica.*

Müller, J. & H.Troschel 1842. *System der Asteriden.* F.Vieweg und Sohn, Braunschweig. 134pp.

O'Connel, M.G., C.B.Alender & E.M.Wood 1974. A fine structure study of venom gland cells in globiferous pedicellariae from *Strongylocentrotus purpuratus* (Stimpson). *J. Morphol.* 142:411-432.

Oldfield, S.C. 1975. Surface fine structure of the globiferous pedicellariae of the regular echinoid *Psammechinus miliaris* (Gmelin). *Cell. Tis. Res.* 162:377-385.

Oldfield, S.C. 1976. The form of globiferous pedicellariae ossicles of the regular echinoid *Psammechinus miliaris* (Gmelin). *Tissue and Cell* 8:93-99.

Pentreath, V.W. & J.L.S.Cobb 1972. Neurobiology of Echinodermata. *Biol. Rev.* 47:363-392.

Pentreath, V.W. & G.A.Cottrell 1971. 'Giant' neurones and neurosecretion in the hyponeural tissue of *Ophiothrix fragilis* Abildgaard. *J. exp. mar. Biol. Ecol.* 6:249-264.

Pérès, J.M. 1950. Sur les pédicellaires glandulaires de *Sphaerechinus granularis* (Lamarck). *Archs Zool. exp. gén.* 86 (Notes et Revues):118-136.

Perrier, E. 1869. Les pédicellaires et les ambulacres des asteries et des oursins. Doctoral thesis, Univ. of Paris. Masson, Paris.

Peters, B.H. 1981. The neurobiology of spines and pedicellariae from *Echinus esculentus* L. Ph.D. Dissert., Univ. London.

Peters, B.H. & G.A.B.Shelton 1981. Electrical activity during a 'simple' behaviour: spine-pointing in a sea-urchin. *Comp. Biochem. Physiol.* 70A:397-403.

Péquignat, E. 1966. Skin digestion and epidermal absorption in irregular and regular urchins. *Nature, Lond.* 210:396-399.

Péquignat, E. 1970. On the biology of *Echinocardium cordatum* (Pennant) of the Seine Estuary. New researches on skin-digestion and epidermal absorption in Echinoidea and Asteroidea. *Forma Functio* 2:121-168.

Prouho, H. 1887. Recherches sur le *Dorocidaris papillata* et quelques autres échinides de la Mediterranée. *Archs. Zool. exp. gén.* (2)5:213-380.

Prouho, H. 1890. Du rôle des pédicellaires gemmiformes chez les oursins. *C.R. Acad. Sci. Paris* 111:62:

Robillard, G.A. 1971. Feeding behaviour and prey capture in an asteroid *Stylasterias forreri. Syesis* 4:191-197.

Romanes, G.J. 1884. Observations on the physiology of Echinodermata. *J. Linn. Soc. (Zool.)* 17:131-137.

Romanes, G.J. 1885. *Jelly-fish, starfish and sea-urchins.* Kegan Paul, Trech., London. 323pp.

Romanes, G.J. & J.C.Ewart 1881. Observations on the locomotor system of Echinodermata. *Phil. Trans. R.Soc.* 172:829-885.

Sars, M. 1846. *Fauna Littoralis Norvegiae*, Heft 1, Christiania.

Sharpey, W. 1836. Echinodermata. In: Todd (ed.), *Cyclopoedia of Anatomy and Physiology*, pp.30-46. London.

Sladen, P. 1880. On a remarkable form of pedicellaria and the functions performed thereby. *Ann. Mag. nat. Hist.* 7:101-114.

Smith, J.E. 1937. On the nervous system of the starfish *Marthasterias glacialis* (L.). *Phil. Trans. R.Soc.* (B)227:111-173.

Smith, J.E. 1965. Echinodermata. In: T.H.Bullock & G.A.Horridge (eds), *Structure and Function in the Nervous Systems of Invertebrates.* pp.1519-1558. W.H.Freeman & Co., San Francisco.

Smith, J.E. 1966. The form and function of the nervous system. In: R.A.Boolootian (ed.), *Physiology of Echinodermata*, pp.503-511. Interscience, New York.

Splechtna, H. & H.Hilgers 1980. Rasterelektronen – und lichtoptisch Unterschungen an den Pedizellarien von *Arbacia lixula* L. In: M.Jangoux (ed.), *Echinoderms: Present and Past*, pp.79-84. Balkema, Rotterdam.

Stephens, G. 1968. Dissolved organic matter as a potential source of nutrition for marine organisms. *Am. Zool.* 8:95-106.
Stewart, C. 1879. On certain organs of the Cidaridae. *Trans. Linn. Soc. Lond.* (2)1:569-572.
Thompson, D'Arcy, W. 1947. *A Glossary of Green Fishes.* Oxford University Press.
Thomson, W. 1874. On the Echinoidea of the Porcupine deep-sea dredging expedition. *Phil. Trans. R. Soc.* 164:719-756.
Uexküll, J.von 1896a. Uber Reflexe bei den Seeigeln. *Z.Biol.* 34:298-318.
Uexküll, J.von 1896b. Vergleichend sinnesphysiologische Untersuchungen II. Der Schatten als Reiz für *Centrostephanus longispinus. Z. Biol.* 34:319-339.
Uexküll, J.von 1899. Die Physiologie der Pedicellarien. *Z. Biol.* 37:334-403.
Uexküll, J.von 1900. Die Physiologie des Seeigelstachels. *Z. Biol.* 39:73-112.
Valentin, G. 1841. Anatomie du genre *Echinus.* In: *Monographies d'Echinodermes vivans et fossiles.* (compiled by L.Agassiz, 1838-1841). Neûchatel.
Weber, W. & M.Grosmann 1977. Ultrastructure of the basiepithelial nerve plexus of the sea-urchin *Centrostephanus longispinus. Cell. Tiss. Res.* 175:551-562.

RECRUITMENT IN ECHINODERMS

THOMAS A. EBERT
Department of Biology, San Diego State University, California, USA

CONTENTS

1 INTRODUCTION

Many, if not all, marine organisms show both temporal and spatial variation in recruitment. Understanding the relative importance of environmental factors that are the proximate causes for the observed variation has been a perennial activity of biologists for at least the past 70 years (e.g. Hjort 1914, Thorson 1950, Frank 1975). The reasons for such investigations include practical application to fisheries (e.g. Iles 1973, Hayman & Tyler 1980) as well as more theoretical considerations of life-history evolution (Murphy 1968, Reaka 1979, Roff 1981). The purpose of this paper is to present information on recruitment in echinoderms and specifically to examine variability in recruitment.

Recruitment means addition of new individuals to a population and may be by immigration or reproduction. Size at recruitment is variable and depends upon the characteristics of the organisms, collecting methods, local conditions, and the particular definition of recruitment that is applied; a new recruit may mean a newly hatched or settled individual or a new reproductive animal or one that is large enough to be taken with a particular sampling device. In this paper I will consider recruitment of individuals to mean that they are just large

Table 1. Recruitment of echinoderm species from offshore bottom communities of the Northumberland coast (from Buchanan 1967).

Class	Species	Year															
		1950					1955					1960					1965
		0	1	2	3	4	5	6	7	8	9	10	11	12	13	14	15
Echinoidea	*Echinocardium cordatum*	E	O	O	O	O	O	O	O	E	O	O	O	O	O	O	E
	Brissopsis lyrifera	●	●	●	●	●	●	●	●	●	●	●	●	E	E	E	E
Ophiuroidea	*Amphiura chiajei*	O	O	O	E	O	O	O	O	O	O	O	O	O	O	O	E
	Amphiura filiformis	●	●	●	●	●	●	●	●	●	●	●	●	E	E	E	E
Holothuroidea	*Cucumaria elongata*	O	O	O	O	O	O	E	O	E	O	O	O	E	O	O	O

● Year classes no longer present in 1965
E Year classes present in 1965
O No recruitment

enough to be collected in proportion to their true abundance. This meaning will distinguish 'recruitment' from 'settlement', which will mean metamorphosis from the plankton, and 'hatching' or 'release', which will be applied to species that are viviparous, brood, or have direct development.

The generally accepted list of environmental factors that cause variation in recruitment is short. Thorson (1950) listed four factors: food, temperature, predation and problems with metamorphosis as affecting the larval stage. There also is evidence that environmental factors affect early survival following settlement.

Thorson (1950) emphasized the role of planktonic predators and maintained that other environmental factors were important only to the extent that they modify the length of time that larvae are at risk in the plankton. Hjort's (1914) 'critical period' concept calls attention to food shortage and starvation of the earliest feeding stage. Coe (1956) stressed the role of currents in bringing larvae into areas where settlement was possible and Peterson (according to Thorson 1950) focused on the importance of early events after settlement in determining recruitment success. Thorson (1950) suggested a role of both positive and negative feedback of adult density on recruitment and this point of view is fundamental to determining fishing potential from stock and recruitment curves (e.g. Cushing & Harris 1973). Although the factors seem reasonable and can be assembled into a plausible model of what causes populations, on the average, to be neither increasing nor decreasing, cause and effect in even the best-studied cases is unclear.

It is still reasonable to ask such fundamental questions as: What determines recruitment density and spatial and temporal variation? Furthermore, because recruitment is part of the total life cycle, it is necessary to ask what role den-

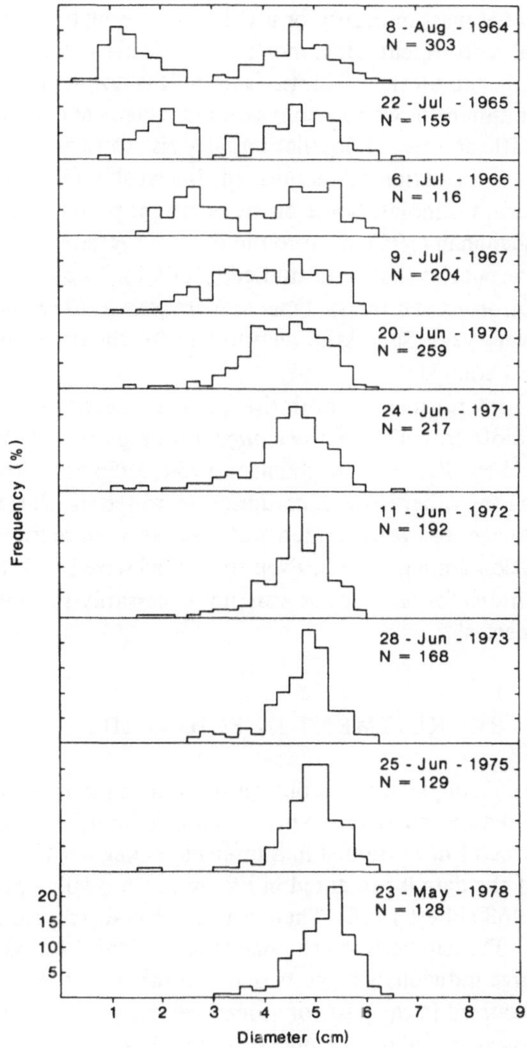

Figure 1. Size-frequency distributions of the echinoid *Strongylocentrotus purpuratus* from 1964 to 1978 in the *Postelsia* zone of Sunset Bay, Oregon (Ebert, unpublished).

sity, and spatial and temporal variation, of one stage play in determining density and variation in recruitment.

There are two fundamental points of view in studies that consider recruitment. The first is that populations are held near average densities by biotic factors and that there is an environmental carrying capacity that is determined by such biotic forces as competition and predation. The second point of view is that average densities are simply statistical averages and that the usual condition of a population may be of two types: usually increasing but with occasional mass mortality or usually decreasing but with occasional highly successful recruitment. Mass mortality or highly successful recruitment could be due either to biotic or abiotic factors. It is not at all clear whether biotic rules are dominant or whether physical and chemical components outside of the homeostatic abilities of populations give rise to variation in recruitment.

Information on recruitment is available for all extant echinoderm classes except crinoids. Some of the recurrent patterns are shown in the work of Buchanan (1967) who found extreme variation in recruitment for three of the five species that were studied (Table 1). Two species, the ophiuroid *Amphiura filiformis* and the spatangoid *Brissopsis lyrifera,* had successful recruitment every year. Successful recruitment for the other species was infrequent and did not coincide.

Recruitment of both the spatangoid *Echinocardium cordatum* and the holothuroid *Cucumaria elongata* took place in 1958 (Table 1). *C.elongata* also had significant recruitment in 1954, 1956 and 1962 but *E.cordatum* did not. On the other hand, *E.cordatum* recruited in 1950 when *C.elongata* did not. Buchanan's work demonstrates variation in recruitment and the lack of correlation among species. Even though all have planktonic larvae, appropriate conditions for one species were not necessarily sufficient for recruitment of other species.

2 RECRUITMENT IN ECHINOIDS

An example of the problems of understanding variation in recruitment is provided by the eastern North Pacific echinoid, *Strongylocentrotus purpuratus,* which I have studied in Sunset Bay, Oregon (43°21′N), since 1963. An example of the data is presented in Figure 1. Unusually good recruitment took place in 1963 (Ebert 1968). There was no or poor recruitment from 1964 through 1978.

The gap between the size modes of the 1963 year class and the mode of large individuals suggests that recruitment of the magnitude of 1963 had not occurred in the past for at least several years, probably back to at least 1959; however, the right shoulder on the first mode in 1964 may indicate moderate recruitment in 1962. Settlement in 1963 was a unique event over a period of at least 20 years in the history of the population of *S.purpuratus* at Sunset Bay.

Recruitment to populations of *Strongylocentrotus purpuratus* is more frequent in southern California and Mexico (Figs.2 and 3). The 1970 size distribution of *S.purpuratus* at False Point (32°48′N) shows a mode at about 1.25 cm diameter (Fig.2), which is composed of individuals that settled in the summer of 1969. In the January 1971 sample, it is apparent that there was poor recruitment in the summer of 1970; however, the presence of large numbers of small individuals in June 1971 shows that there was recruitment in the spring of 1971.

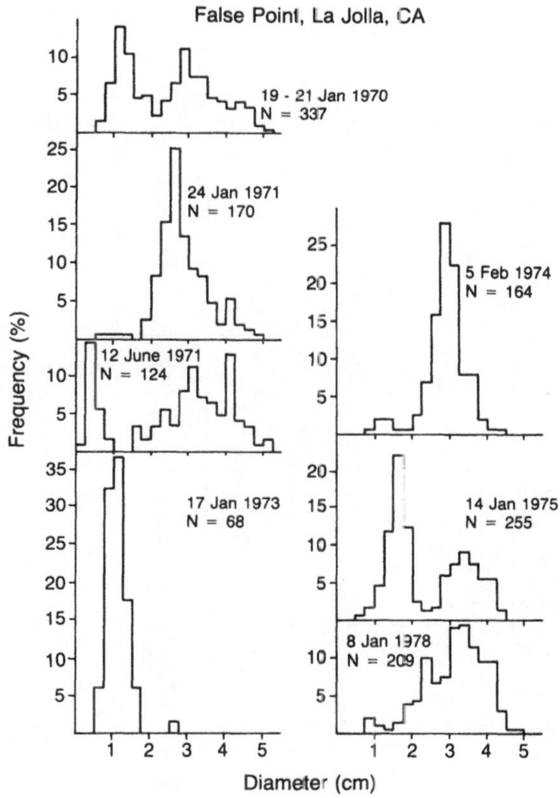

Figure 2. Size-frequency distributions of *Strongylocentrotus purpuratus* from 1970 to 1978 at False Point, La Jolla, California (Ebert, unpublished).

In late summer of 1971 surface sea water exceeded 24°C. The lethal limit for *S.purpuratus* is about 24°C (Farmanfarmaian & Giese 1963), but individuals cannot tolerate temperatures consistently ≥22°C (Ford *et al.* 1976). There was a mass mortality at False Point possibly after summer settlement because there were very few individuals at False Point during fall 1971 and also during winter 1972 (Ebert, unpublished). Substantial recruitment took place during the summer as shown by the large number of small individuals found in January 1973. Recruitment was poor during the summer of 1973, good in 1974 and poor in 1977. The position of a mode at about 2.4 cm diameter in 1978 suggests that there was substantial recruitment in 1976.

The data from Papalote Bay, Baja California, Mexico (31°45'N) show substantial recruitment every year from 1962 through 1969 (Fig.3). Recruitment at Papalote Bay takes place more than just once per year as shown by the presence of a mode less than 1.0 cm diameter in distributions from January, April and June through November. There is year-to-year variation; a large mode of

Papalote Bay, Baja California, Mexico

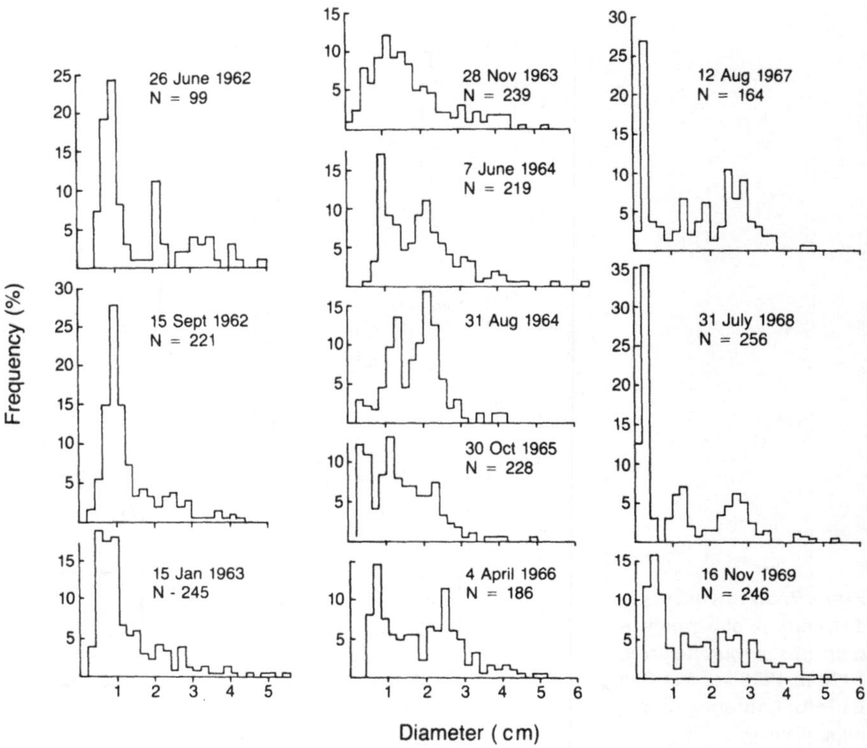

Figure 3. Size distributions of *Strongylocentrotus purpuratus* at Papalote Bay, Baja, California, Mexico (from Pearse *et al.* 1970).

small individuals is shown in the August 1967 distribution but there are very few small individuals shown in the August 1964 graph. Similarly, the November 1963 and 1969 distributions reflect differences in recruitment.

A latitudinal cline of recruitment is shown by another eastern North Pacific echinoid. In British Columbia, Canada, recruitment of *Strongylocentrotus franciscanus* is infrequent and many populations consist solely of large individuals (Bernard & Miller 1973, Low 1975). In southern California and Mexico, recruitment is frequent and may take place nearly every month of the year (Fariás Sánchez 1980, Tegner & Dayton 1981). The difference in recruitment success between northern and southern populations is striking and it is reasonable to ask: What is the basis for the correlation between latitude and recruitment?

Development rates of marine invertebrates tend to be modified by temperature. According to Kinne (1970) it would be reasonable to expect development

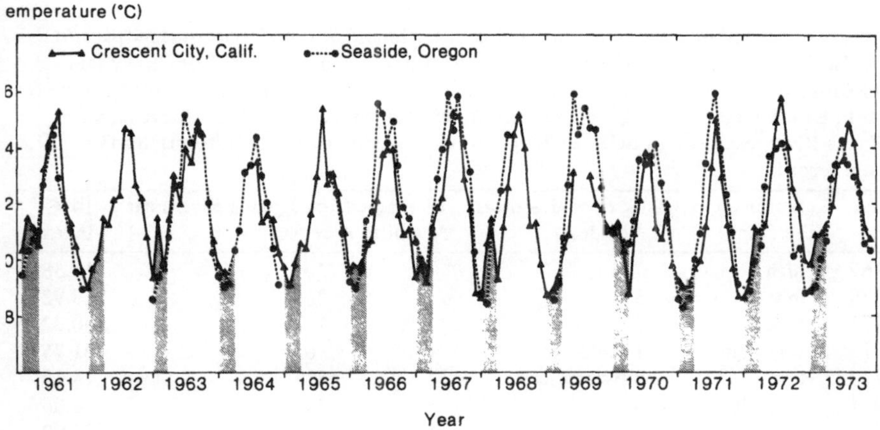

Figure 4. Average monthly sea surface temperatures in northern California and Oregon from 1961 through 1973. Stippled area gives the approximate breeding season of *Strongylocentrotus purpuratus.*

to be positively correlated with temperature; although culture at ambient temperatures along a latitudinal gradient has, in some cases, resulted in similar rates (cf. Dehnel 1955).

According to Thorson's (1950) view of the importance of predation in the plankton, it would be reasonable to predict that the faster the rate of development, and hence passage through the planktonic stage, the greater the settlement. Although a second-order relationship between adult growth and temperature is given for *S.purpuratus* by Pearse *et al.* (1970) with a maximum growth of 14°C, larval development, at least at the earliest stages, does not have an optimum temperature (Tyler 1936). Development rate is a monotonically increasing function of temperature.

Results for echinoid recruitment that are consistent with Thorson's temperature hypothesis are presented by Kawana (1956 cited by Matsui 1966) who showed that echinoid recruitment was favored by warm temperatures of the Tsushima current. The exact cause of increased recruitment, however, is not certain because Kawana also showed that densities of juvenile herring were negatively correlated with warm conditions and hence the effect of warm water may be due to decreased predation rates rather than due to changes in development times, or due to some other covariate of temperature.

The temperature of the northern California and Oregon coasts during the winter of 1962-63 was not unusual (Fig.4). Recruitment was slightly higher in 1971 than in most other years and the winter of 1970-71 was somewhat warmer than average. There does not seem to be any obvious association between recruitment of *S.purpuratus* and warm winter temperatures off Oregon.

There is year-to-year variation in winter and spring temperatures off south-

Table 2. Relationship between recruitment of the echinoid *Strongylocentrotus purpuratus* at False Point, California, and cold water temperatures in winter, runoff from heavy rains and Barilotti's 14°-index (the percent of the time that the temperature at 5 m off Scripps Pier in La Jolla less than 14°C; Winter temperatures shown in Figure 5; discharge data for Tijuana River taken from USGS 1969-1977 water resources data for California; Surface water records.

Year	Recruitment at False Point	Rank of cold winters (1 = coldest)	Discharge from Tijuana River near Nestor in Acre feet	14°C Index
1969	high	7	3750	17.58
1970	low	5	35	13.92
1971	high	2	12	30.72
1972	very high	1 (coldest)	0	31.75
1973	low	8 (warmest)	81	4.12
1974	very high	3	54	36.86
1975	?	–	50	34.90
1976	high	4	663	29.90
1977	low	6	96	5.93

Association of cold winters and recruitment

		Rank 1-4	Rank 5-8
Recruitment	High	4	1
	Low	0	3

Association of 14°C index and recruitment

		Index <15	Index ⩾15
Recruitment	High	0	5
	Low	3	0

ern California (Fig.5). Recruitment is not independent of winter temperature (Table 2); however, the relationship is opposite to that predicted by Thorson (1950) and Kinne (1970): recruitment was greatest during the coldest winters when temperature would have depressed development rates of larvae.

Freshwater runoff has been correlated with recruitment in invertebrates and fish (e.g. Sutcliffe 1973) and with the population increases in the crown-of-thorns asteroid *Acanthaster planci* (Birkeland, 1982). These investigators suggest that runoff increases nutrients in near-shore areas, which cause phytoplankton blooms, and which, in turn, increase the rate of larval development and settlement success. An alternative interpretation is that phytoplankton blooms synchronize spawning and so improve reproductive success (Himmelman 1980). The implication of these studies is that phytoplankton levels usually are so low that spawning success is decreased or larvae starve or have such prolonged development times that predation in the plankton is complete.

There is no correlation between runoff from the Tijuana River in South San

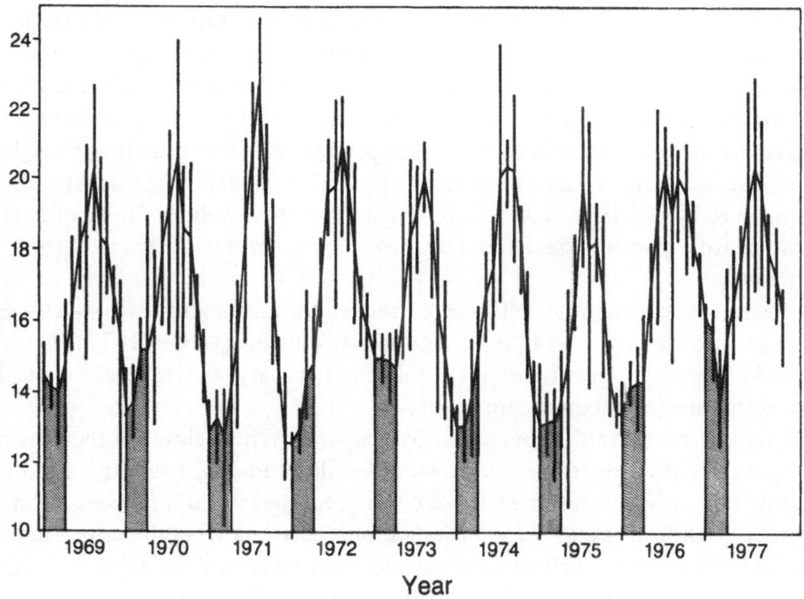

Figure 5. Average monthly sea surface temperatures off Scripps Pier, La Jolla, California from 1969 through 1977. Vertical lines give monthly temperature ranges and stippled area shows approximate breeding season.

Diego and recruitment at False Point (Table 2). In 1969 there was extensive flooding in San Diego and substantial recruitment. There also was excellent recruitment in 1972 when there was no measurable runoff from the Tijuana River. The major contribution of both freshwater and nutrients to the nearshore environment of San Diego, however, is not from runoff but from sewage discharge off Point Loma, about 16 km to the south of False Point. Discharge has steadily increased over the past several decades and so year-to-year variation in echinoid recruitment at False Point is independent of this source of nutrients.

In southern California, the 14°C isotherm approximates the depth of the thermocline or bottom of the nutrient-poor layer (Cairns 1968, Strickland 1970, Kamykowski 1972). An upward movement of the 14°C isotherm, and hence the thermocline, generally is associated with upwelling (Smith 1968) although temperatures in the California current are lowered by advection of northern water to the south (Chelton 1981). An index based on the depth of the 14°C isotherm developed by C.Barilotti (personal communication) shows a positive association with recruitment (Table 2). Years from 1969-1977 when there was strong upwelling, based on Barilotti's 14°C-index, were years with low surface temperatures in winter and high *S.purpuratus* recruitment at False Point.

There are problems with the correlation between recruitment and the 14°C-index in terms of the possible mechanisms by which upwelling could operate to promote recruitment. Periods of strong upwelling correspond with periods of northerly winds when Ekman transport moves water offshore (Smith 1968, Bakun 1973). Upwelling brings nutrient-rich waters into the zone of mixing above the thermocline. It is the increased productivity that usually is thought to be the cause of increased recruitment of marine animals (e.g. Cushing 1971, Hayman & Tyler 1980). However, during periods of upwelling, larvae of near-shore benthic invertebrates should be moved away from potential sites for settlement.

Recruitment of another echinoid correlates with upwelling. *Encope stokesi* recruits throughout the year at Venado Beach, Panama (Dexter 1977). The period of peak settlement, however, is during February, the dry season, which also is the time of maximum upwelling.

Larvae may return to shore either by deeper currents below the thermocline or by a shift of winds to the south, which would cause negative Ekman transport of the surface waters back toward shore. Settlement of *S.purpuratus* in southern California takes place during summer just following the usual annual period of strongest upwelling as calculated from pressure fields (Bakun 1973) but nearly five months after the maximum values of Barilotti's 14°C-index.

Water movement in Oregon was very low during the first two quarters of 1962 and 1963 (Bakun 1973). Ekman transport values showed positive anomalies for the first quarter and negative anomalies for the second quarter. Usual transport is negative in the first and positive in the second quarters, so the anomalies indicate unusually weak onshore and offshore water movement. The two years had unusually calm winds and low storm frequency during the first quarter of each year (Hayman & Tyler 1980) when larval development takes place. In October 1962 a severe storm struck the Oregon coast but it was followed by unusually calm conditions during January-March 1963 (Hayman & Tyler 1980). Unlike *S.purpuratus* in southern California, unusually good recruitment of this species in Oregon seems to be associated with weak upwelling.

The lack of a consistent relationship between strong upwelling and recruitment of *S.purpuratus* in southern California and Oregon is similar to the relationship between upwelling and zooplankton development shown by Chelton (1981). A significant positive correlation exists in southern California but the relationship breaks down farther north. Based on salinity patterns, Chelton suggested that southward advection of water in the California Current rather than upwelling may explain zooplankton variation.

Transport from the north may be the correct explanation for increased recruitment rates of *Strongylocentrotus* spp. in the south. The dominant direction of the California Current is north to south. As southern populations are 'down stream' from more northern populations, there may be more larvae available for settlement at southern sites.

Strongylocentrotus droebachiensis is circumboreal and in North America ranges from Washington north through the Aleutian Islands. The recruitment of *S.droebachiensis* is predictable in a geographic region that shows variable recruitment for *S.purpuratus* (Duggins 1980), which is similar to the recruitment patterns (Table 1) reported by Buchanan (1967). As both species have planktotrophic larvae, it seems unlikely that food density is the cause of the variation or that planktonic predators discriminate between the two species. There may, however, be different food requirements as has been shown for larvae of *Lytechinus pictus* vs. *Arbacia punctulata* (Hinegardner & Tuzzi 1981).

Strongylocentrotus droebachiensis may develop more rapidly than *S.purpuratus* (Strathmann 1978) and this difference may be the proximate cause of differences in recruitment success. Minimum times from fertilization to settlement is 51 days for *S.droebachiensis* and between 63 and 86 days for *S.purpuratus.* The maximum for *S.droebachiensis* is 152 days but the maximum is unknown for *S.purpuratus.* It also is not known if *S.droebachiensis* can live longer in the plankton than *S.purpuratus* and still settle.

Recruitment of *Strongylocentrotus purpuratus* varies with intertidal position. At Sunset Bay, Oregon, based on the percent contribution of the 1963 year class, there were more recruits high in the intertidal than at the lowest stations (Table 3). The differences would seem most easily explained by behavior of the larvae at the time of settlement. A reasonable hypothesis is that the final stimulus to settle is change in water motion from turbulent to calm as waves carry larvae into pools during a falling tide; however, I know of no studies that have examined this aspect of behavior.

Other factors associated with the recruitment of *Strongylocentrotus purpuratus* concern events at or near the time of settlement. Significant mortality may occur during settlement and within the first few weeks of benthic life. Recruitment is better outside kelp beds *(Macrocystis pyrifera)* compared with inside (Pearse *et al.* 1970, Tegner & Dayton 1981). Small predators such as hydroids, bryozoans or serpulid polychaetes (Bernstein & Jung 1979) on the kelp blades or small fish that use the kelp as habitat (Hobson & Chess 1976) may eat the larvae before they settle. Alternatively, the kelp may modify water movement so that fewer larvae may enter the bed or a change in water movement may stimulate settlement at the edge of the bed.

Site selection may play a role in causing spatial variation. Recruitment appears to be better around or under adults (Moore *et al.* 1963, Ebert 1968, Tegner & Dayton 1977, Schroeter 1978). Cameron & Schroeter (1980) have shown that there are at least two interpretations for the association between young and adult echinoids. As there was no increased settlement in the presence of adults or substrates where adults had been, they suggested that either there is better survivorship of young that settled under adults or that larvae move and accumulate under adults.

Highsmith (1977) argued that predation by tube-building tanaidacean crus-

Table 3. Relative frequency of the 1963 year class in 1965 of *Strongylocentrotus purpuratus* along an intertidal cline in Sunset Bay, Oregon. The 1963 year class was separated from size distributions using the program developed by MacDonald & Pitcher (1979). Names of areas follow Ebert (1968); rank correlation coefficient = 0.83.

Area	Elevation relative to datum (MLLW) in meters	Fraction of the population composed of the 1963 year class ± SE
Above *Postelsia* Zone and North	2.5 m	0.88 ± 0.02
Postelsia Zone	0.9	0.39 ± 0.04
Eel Grass Area	0.7	0.61 ± 0.03
Boulder Field	−0.2	0.42 ± 0.04
Below *Postelsia* Zone and South	−0.5	0.19 ± 0.04
Below *Postelsia* Zone and North	−0.6	0.15 ± 0.03

taceans on newly-settled sand dollars *(Dendraster excentricus)* may explain local patchiness of adult distribution. This observation is consistent with the apparent preference of larvae for sand with adults (Birkeland & Mesmer 1978), which also shows low tanaid densities. The positive relationship between recruitment and adult densities of *D.excentricus* is contrary to the observations of Timko (1979) who interpreted her data to show that adult density had a negative influence on recruitment. Chia & Atwood (1982) however, were unable to confirm Timko's results on larviphagy.

Both spatial and temporal variation in recruitment is shown by many other echinoid species. On a local scale (1 km), Fuji & Kawamura (1970) showed differences in subtidal recruitment density of *Strongylocentrotus intermedius* at Ikantai in southern Hokkaido. The highest concentration of newly settled animals was on a cobble (<30 cm diameter) bottom and adults were found in habitat with boulders (>50 cm).

The importance of local substrate condition also is suggested by Lang & Mann (1976) for *S.droebachiensis.* Based on age distributions determined from natural growth lines, there was increased settlement one year following loss of kelp. The mechanism for this response is not clear. It may be that physical change (loss of kelp) of the substrate makes it more attractive to competent larvae or it is a manifestation of the same phenomenon that decreases recruitment of *S.franciscanus* inside kelp beds (Pearse *et al.* 1970, Tegner & Dayton 1981). It is also possible that the data do not represent the actual age structure of the populations. These are problems with using 'growth lines' to determine age (Pearse & Pearse 1975) and their validity has yet to be established for *S. droebachiensis.*

Small scale variation in recruitment also has been shown for *Lytechinus variegatus* in Puerto Rico (Rivera 1979). Three out of the four study sites in Jobos Bay showed recruitment in 1975-76. Length of the recruitment period

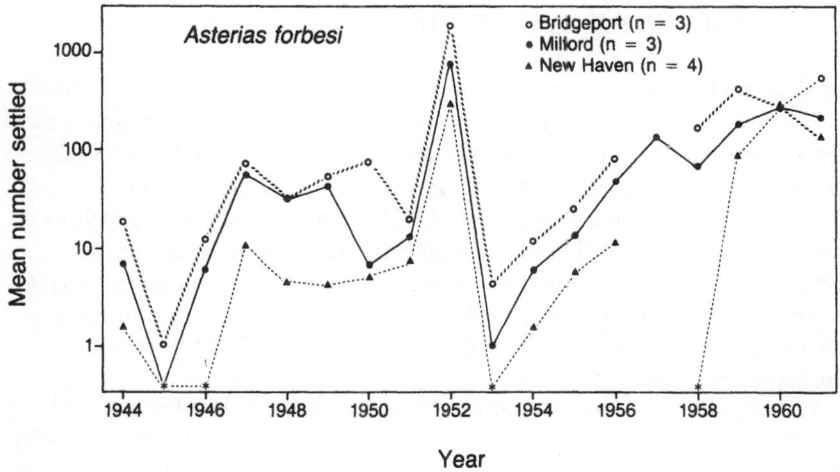

Figure 6. Mean settlement of asteroid, *Asterias forbesi,* per 100 oyster shell surfaces along the Connecticut coast of Long Island Sound from 1944 to 1960 (from Loosanoff 1964).

was not the same for these three sites and one site (station #1) had two recruitment episodes. The site with no recruitment (station #4) and station #1 were about 6 km apart and at opposite ends of Jobos Bay. Over a two year period (1974-76) station #4 showed no recruitment and station #1 had the best recruitment. Intermediate stations had intermediate levels of recruitment for both years.

The pattern of year-to-year variation in recruitment in echinoids frequently includes exceptionally good years as well as years with no apparent recruitment. In addition to work already cited, extreme variation that includes complete failure for some years has been reported for *Strongylocentrotus droebachiensis* (Foreman 1977) in British Columbia, *Paracentrotus lividus* (Ebling *et al.* 1966) in Loch Ine, *Lytechinus variegatus* (Moore *et al.* 1963, Camp *et al.* 1973) in Florida, *Mellita grantii* and *Encope grandis* (Ebert & Dexter 1975) in the Gulf of California and *Mellita quinquiesperforata* (Lane & Lawrence 1980) in Florida.

Events in the plankton, at the time of settlement, and during the early benthic stages are not understood in sufficient detail to evaluate their relative importance as critical stages in echinoid life cycles. The relationship between recruitment of *Strongylocentrotus purpuratus* and Barilotti's 14°C-index or Bakun's upwelling index suggest that availability of larvae to settle may be determined by water movement (cf. Jackson & Strathmann 1981) but details of settlement may be influenced by local factors such as bathymetric position and presence of adults and small predators.

The field work done on *Asterias forbesi* on the Atlantic coast of the US, summarized by Loosanoff (1964), provides an excellent focus for discussing recruitment in asteroids. Loosanoff (1964) summarized work done in Long Island Sound from 1944 to 1961 at three sampling locations along the Connecticut coast. Settlement was measured twice weekly at three depths from late June to October by counting small individuals taken in oyster-shell collectors.

Settlement of *A.forbesi* varied from year to year over four orders of magnitude from virtually zero to over 1000 per one hundred oyster-shell surfaces (Fig.6). Settlement consistently tended to be greatest at Bridgeport, the southernmost station.

An analysis of covariance of Loosanoff's data (Table 4) is consistent with his conclusions that settlement declines from south to north and increases with depth over the range of depths that were examined. To perform the analysis, I equalized sample number by taking the average of two 10-foot samples at Milford (other sites and depths were represented by a single sample) and stabilized variance by using a logarithmic transformation of the data (log (set + 1.0)).

The adult population did not fluctuate to the same degree shown by settling individuals (Fig.7). Moreover, heavy settlement did not follow the greatest adult densities, contrary to the analysis of Burkenroad (1957), nor did heavy settlement necessarily lead to an increase in the adult population. Heavy settlement in 1941 was associated with an apparent drop in adult density from 1941 to 1942. On the other hand, heavy settlement in 1952 is associated with a rise in the adult population from 1952 to 1953. The two important features are the stability of the adult population and the lack of tight coupling between either adult density and settlement density or settlement and subsequent adult density.

Consistent differences in settlement of *A.forbesi* among sites that are close

Table 4. Analysis of settlement of the asteroid *Asterias forbesi* at three sites in Connecticut from 1944 through 1961. Data taken from Loosanoff (1964). Grouping was done by year; the number of asteroid set is the dependent variable; and the three sites and three depths are coavriates. The number set was transformed using set = log (set + 1.0); analysis done using BIOMED Program BMDP2V (Dixon & Brown 1979).

Source	Sum of squares	Degrees of freedom	Mean square	F-value	Slope (β) estimates
Mean	6.806	1	6.806	129.80*	–
Year	43.953	16	2.747	52.39*	–
Sites	3.052	1	3.052	58.21*	–0.173
Depth	1.256	1	1.256	23.95*	0.111
Both covariates	4.301	2	2.154	41.08*	–
Error	7.027	134	0.052	–	–

* Significant at $p < 0.01$

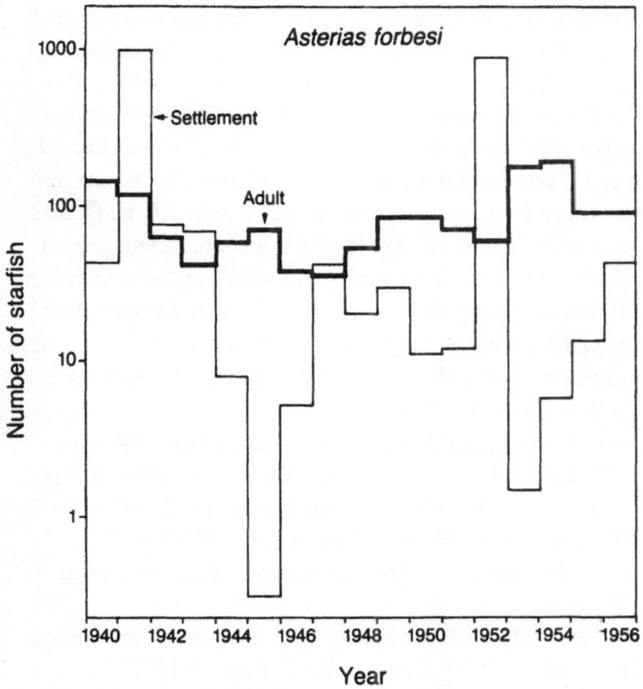

Figure 7. Settlement density (average of all sites and depths) and adult density (number taken over a 10,000 foot dredge course) of *Asterias forbesi* in Long Island Sound (from Loosanoff 1964).

together is similar to the recruitment pattern shown by *Pisaster ochraceus* on San Juan Island, Washington (Menge & Menge 1974) where Point Caution consistently had good recruitment and Gull Reef did not.

Because settlement was measured on collectors for *A.forbesi*, it is possible to argue against biotic differences among sites. The same substrates were used to sample settlement at all sites and remained in the field for only a few days so that build-up of small predators that would eat young individuals is unlikely. It is more reasonable to suggest that the data show differences in supply of larvae rather than differences in early events following settlement or differences in substrate. Two important factors would be differences in water movement (currents) or differences in predation in the plankton. The consistent year-to-year pattern suggests that predation is less likely a factor than water movement.

Biotic factors cannot be ruled out for the differences in recruitment shown by *P.ochraceus* (Menge & Menge 1974). Nothing is known of settlement and early survivorship and the observed differences in recruitment could be settlement differences, as in *Asterias forbesi*, or could be differences in the distribution of small predators or suitable food for newly metamorphosed *P.ochraceus*.

Birkeland (1974) has argued that asteroids that are food generalists, such as *Mediaster aequalis,* tend to show more regular recruitment than specialists, such as *Hippasteria spinosa.* Adult *Pisaster ochraceus* eat a wide variety of foods (Mauzey *et al.* 1968) and so, based on Birkeland's (1974) arguments, newly settled individuals should not have dietary restrictions that would lead to increased mortality if a single food item was rare. The argument is complicated, however, because there are examples such as *Acanthaster planci* (Lucas 1974) and *Stichaster australis* (Barker 1977) for which diets of newly settled individuals are very different from those of adults and where settlement seems to be restricted to areas with appropriate food items. A further complicating example is *Mediaster aequalis* which shows preference for a particular substrate, tubes of *Phyllochaetopterus prolifica* (Birkeland *et al.* 1971), but feeds on a wide variety of animals that live on the tubes.

Acanthaster planci had a rapid population increase during the 1960's on the Great Barrier Reef (Endean 1973, 1976), Guam (Chesher 1969) and other scattered localities throughout the Indo-West Pacific (Endean & Chesher 1973, Weber & Woodhead 1970, Sale *et al.* 1976) and the Red Sea (Campbell & Ormand 1970). The review by Potts (in press) provides an excellent critical review of the data.

A controversy exists over whether these events were unique and associated with human activity (e.g. Endean 1973, Chesher 1969, Randall 1972) or represent a major though not unique event within the normal range of reef function and not primarily anthropogenic (e.g. Dana, Newman & Fager 1972, Vine 1973).

Some evidence based on sediment analysis suggests that the rapid increase was an unusual though not unique event (Frankel 1975, 1977a, 1977b) with hundreds of years separating such population increases. The populations declined (died? migrated?) during the 1970's but there is evidence that there have been new recruitment episodes in Palau (Birkeland 1979), Siapan (Birkeland, personal communication), Okinawa (Endean, personal communication) and on the Great Barrier Reef (Cameron & Endean 1981).

Endean (1974) and Endean & Stablum (1973) argued that predation on adults regulates populations and that the decline in certain predatory snails *(Charonia tritonis)* or such fish as groupers led to the population increases of *Acanthaster.* Their view is that variation in adult numbers causes variation in recruitment, basically the same argument as that of Vine (1973). Vine's particular interest involved the relationship between the percent fertilization and adult density. The rate at which the probability of fertilization rises as density increases would determine the importance of aggregation for *A.planci.* No studies have attempted to measure fertilization as a function of density for any marine animal.

If *A.planci* is long-lived, as *Asterias forbesi* seems to be, it is reasonable to question whether adult density should determine recruitment density. It does not seem to be true for *A.forbesi* (Fig.7) although *A.forbesi* forms aggregations (Gemmill 1914, Galtsoff & Loosanoff 1939) which may involve spawning but more likely are associated with feeding (Sloan 1980).

Ormond & Campbell (1974) proposed a positive feedback loop in which adults eat coral, which provides more suitable settling space for larvae. More space leads to more larvae settling, which results in more adults. However, larvae do not settle on living coral (Yamaguchi 1973a, 1973b, Lucas 1973, 1974), but on encrusting coralline algae such as *Porolithon.* Metamorphosis and settlement, however, do seem to be a particularly critical period in the life of an asteroid. In one of Yamaguchi's cultures (Yamaguchi 1973a), only 57 of 120 fully developed brachiolariae of *A.planci* that metamorphosed were recovered as juveniles from coral debris. The fate of the young juveniles is not known, but Yamaguchi speculated about small predators.

Kenchington's (1977) analysis of populations of *A.planci* on the Great Barrier Reef is consistent, in part, with Endean's hypothesis, in that he suggested a positive relationship between adult density and recruitment. Kenchington (1977) analyzed a large number of size distributions and attempted to deduce growth and recruitment. He made two assumptions crucial to his arguments: 1) Modes in size distributions are year classes that usually are just one year apart; 2) Growth characteristics are very similar in all populations of *A.planci* along the Great Barrier Reef. The number of year classes that Kenchington extracted from the size distributions range from one to four. Twelve of the distributions had one year-class, 20 had two classes, 13 had three, and one had four.

A great difficulty with examining published size distributions of *A.planci* is that there has been no continuity in sampling. The data of Laxton (1974) and Endean & Stablum (1973) are for various sites with single samples. It is unwise to evaluate change with such data. Kenchington (1977) used these distributions together with unpublished data on the sizes of animals used for gonad studies by Pearson (Pearson & Endean 1969) and Lucas (1973). Pearson & Endean stated that adult individuals were collected and, because the purpose was gonad analysis, the same is implied by Lucas. Such samples are not suitable for size analysis. There are no sequences of size distributions that can be used to examine variation in recruitment at single locations. Because data that exist are not suitable for Kenchington's analysis, it is not clear whether there were multiple recruitment episodes or not. In particular, it is not clear whether there was just one or a few successful years for recruitment and whether the successful years might have been the same years for the entire reef.

Adult populations in Kenchington's model go to local extinction very rapidly within a few years after recruitment. This in itself is of considerable interest because I know of no data that support the view that echinoderms ever starve to death. This statement includes a consideration of the arguments (and lack of data) of North (1974) on the role of starvation in limiting echinoid populations in southern California following the disappearance of the sea otter. Adult *A.planci,* according to Kenchington, must be short lived rather than long lived. However, Cameron & Endean (1981) presented arguments for believing that *A.planci* is long lived; specifically, this species is venomous (the only asteroid species that is), has a spiny armature, virtually no animal parasites, possesses toxic tissue saponins, and can utilize the wax reserves of coral (cetyl palmitate). There are, however, no measurements of survival in the field that indicate longevity in *A.planci.*

Size distributions of *A.planci* sometimes have more than one mode (Laxton 1974) though most are unimodal or irregular. Although the presence of a single mode does not necessarily mean one age-class, the presence of two modes does indicate at least two classes. 'Class' does not have to mean 'year class' and can result from local environmental effects. The data of Laxton (1974) show not only more than one recruitment, but also what Laxton interpreted as settlement near the edge of the reef and subsequent movement. Such an interpretation of his data seems reasonable but differential survival and an associated sampling problem could explain them. The density of *A.planci* is low at reef crests where small individuals are found and increases in more protected areas where large individuals are found. Collection of a sample of 100 individuals must be done by searching a greater area on the reef crest to obtain a sample than on the slope. Hence even with similar densities of small individuals on the crest and on the slope, there is a greater chance for them to appear in the reef crest samples because greater areas were sampled to obtain sufficient numbers for a size distribution. Existing data are not sufficient to decide whether this is the situation or that settlement takes place at the reef crest and individuals move down into calmer waters as they grow. Replicate sampling of crests and slopes are necessary to determine whether densities of small individuals really differ.

A.planci may be long lived in the more temperate areas such as Hawaii and short lived at the northern end of the Great Barrier Reef and in Palau and, like *Strongylocentrotus purpuratus,* may show changes in recruitment with latitude. Recruitment of *A.planci* does appear less predictable in the temperate areas (Laxton 1974) of the Great Barrier Reef than in the more tropical areas and an unusual recruitment event is a reasonable interpretation of the data of Branham *et al.* (1971) as suggested by Ebert (1973) and Yamaguchi (1975b).

If *A.planci* is longer lived in more temperate parts of its range and possibly slower growing, a few major and simultaneous recruitment episodes along the entire reef would show a more rapid build-up at lower latitudes (faster growth) and more rapid decline (higher mortality or more movement). There would be an apparent shift in the population center from north to south simply because animals were developing more slowly in the south and living longer. Endean's migration hypothesis cannot be ruled out as Kenchington tried to do. Migration is an alternative to spread by recruitment and does fit with rapid build-up and decline of populations of *A.planci;* however, it must be made very clear that migration is not the only alternative to Kenchington's model.

Asterias forbesi, Pisaster ochraceus and *Acanthaster planci* have planktonic larvae, but a wide variety of reproductive methods are shown by other asteroids. There are few studies that have focused on the consequences of these differences in terms of recruitment success and so, clearly, it is a fruitful area of research.

Asterina gibbosa and *Patirella exigua* have direct development without a planktonic stage and females produce unguarded egg masses (Crump & Emson 1978, Lawson-Kerr & Anderson 1978). *Asterina phylactica* produces egg masses, which it broods (Emson & Crump 1979), as does *Leptasterias hexactis* (Chia 1966). *Patiriella vivipara* (Dartnall 1969) and *P.pseudoexigua pacifica* (Oguro *et al.* 1978) are viviparous: development is internal and juveniles escape through the female's body wall. Asexual reproduction occurs as fission (e.g. *Nepanthia belcheri*) or by regeneration of entire animals from arm parts as shown by *Linkia multifora* (Emson & Wilkie 1980). This list of species is by no means exhaustive but is intended to show the range of reproductive methods that must be considered in terms of modifying recruitment success.

Increased investment in individual offspring should improve recruitment success and the few data that are available for asteroids support this contention. Improved recruitment success means that an individual should require fewer reproductive episodes to replace itself in a population. Brooding should make recruitment more predictable than broadcasting and so, using the logic of bet-hedging (Murphy 1968, Stearns 1976), brooders should be shorter lived than broadcasters. This comparison has been made for *Pisaster ochraceus* (a broadcaster) and *Leptasterias hexactis* (a brooder) by Menge (1975) who concluded the *P.ochraceus* lives about three times longer than *L.hexactis.* Based on this analysis, *Leptasterias hexactis* does appear to have more successful annual recruitment than *P.ochraceus.*

A similar comparison has been made between *Asterina gibbosa* (non-brooder) and *A.phylactica* (brooder) by Emson & Crump (1979). By their estimates, *A. phylactica* has a mortality rate about twice as great as *A.gibbosa.* Again, this suggests that brooding improves recruitment success; however, a final complication is that there is a possible latitudinal effect. *A.gibbosa* grows faster, matures earlier and may be shorter lived in the Mediterranean than it is in Britain (Crump & Emson 1978).

Do all brooders have more successful recruitment than broadcasters? Are viviparous species such as *Patiriella vivipara* more successful than similar species that produce unguarded egg masses, such as *P.exigua*? Are asteroids that reproduce by fission the most successful and also the shortest lived?

The answers to these questions may not be simple because reproduction does not exist outside the total set of adaptations and associated attributes (Stearns 1977, Gould & Lewontin 1979). Yamaguchi (1975a) provided a list of asteroid

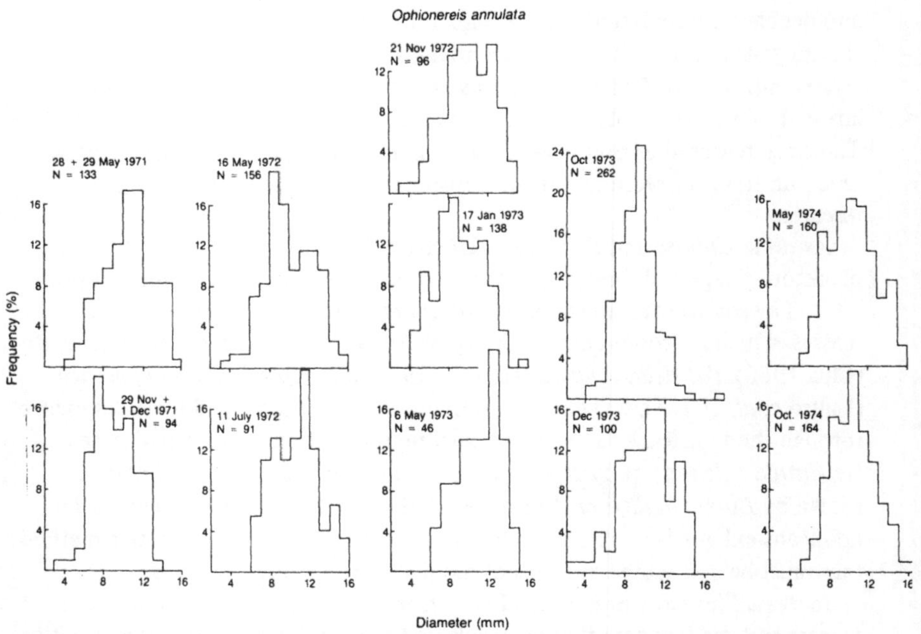

Figure 8. Size frequency distributions of the brittlestar *Ophionereis annulata* in the intertidal of False Point, La Jolla, California, from 1971-1974. Data for Oct.1973-Oct.1974 taken from Muscat (1975).

species in Guam with their relative commonness and mode(s) of reproduction. If relative fitness is measured by abundance, there is no relationship between fitness and mode of reproduction. There are species with planktotrophic larvae that are very common and others that are very rare. The same is true for species that are capable of fission; however, all such species also appear to have planktonic larvae (Emson & Wilkie 1980). If one takes a more existential approach (Slobodkin & Rapoport 1974), they are all equally fit because they continue to exist in Guam and it may not be any better to exist at high densities than at low densities, though it would seem that large populations are less likely to go to extinction than small populations (MacArthur & Wilson 1967), all other things being equal.

4 RECRUITMENT IN OPHIUROIDS

Hendler (1975) has evaluated the adaptive significance of developmental patterns in ophiuroids: planktotrophic, direct development (i.e. viviparous) and abbreviated (lecithotrophic and demersal). Newly released juveniles of viviparous species are larger than newly settled juveniles of planktotrophic species and the larger size probably is advantageous. A resonable extension

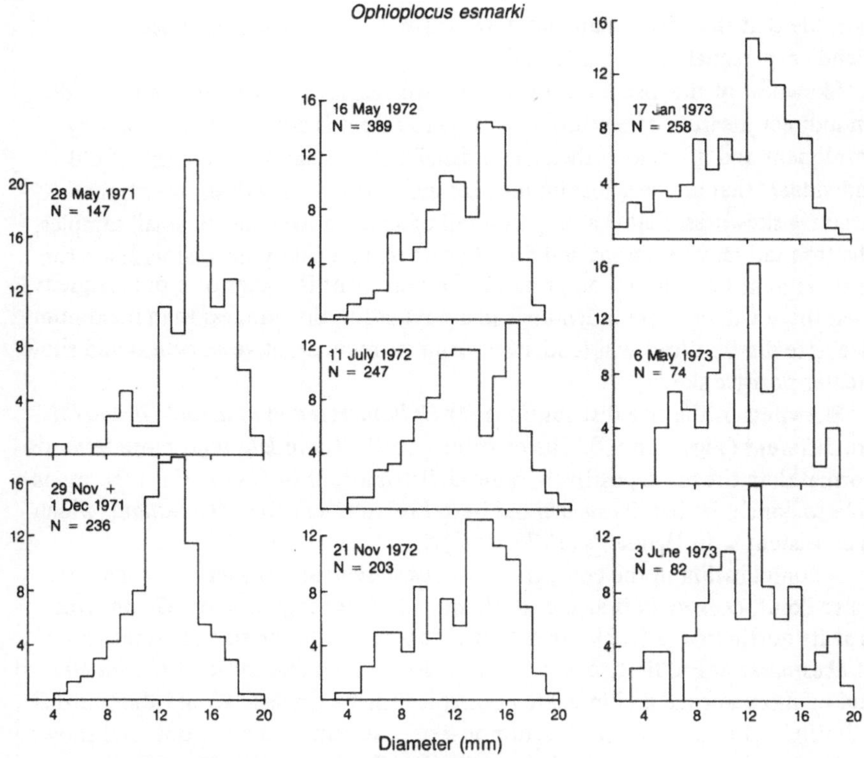

Figure 9. Size distributions of the brittlestar *Ophioplocus esmarki* at False Point, California, from 1971-1973 (Ebert, unpublished).

of this idea, as was argued for asteroids, is that viviparous species should have more predictable recruitment. It should be possible to see differences in recruitment success in size distributions.

Two of the most abundant species at False Point, La Jolla, California, are *Ophionereis annulata,* which has a vitillaria larva (G.Hendler, 1982) and so has abbreviated planktonic development, and *Ophioplocus esmarki,* which is viviparous, in Hendler's (1975) use of the word or is a brooder according to Chia (1978). These two species can be compared to test Hendler's prediction that brooders should have more predictable recruitment than species with planktonic larvae. Size distributions of *O.annulata* should show fewer small animals than distributions of *P.esmarki.*

Based on Hendler (1975, Table 3), newly settled *O.annulata* should be between 0.2 and 0.5 mm. Newly released *O.esmarki* are about 0.5 mm (S.Rumrill, personal communication). Very small individuals are not evident in the distributions of either species (Figs.8 & 9) though they were specifically searched for (Muscat 1975). However, searching was done only under rocks and it is possible that the appropriate habitat, such as tops of rocks, was missed (G. Hendler, personal communication).

Skewness of the distributions for the two species are different and provide an indirect measure of recruitment. If recruitment is either very low or very infrequent and life is long then a size distribution usually will consist of old individuals that are near maximum size; the distribution will show extreme negative skewness, that is a very thin tail of small individuals. In small samples, the thin tail may be absent and distribution will be nearly normal with a mean at maximum size. On the other hand, if recruitment is higher or more frequent then there will be more individuals in a population that are less than maximum size, size distributions will tend away from extreme negative skewness and show greater positive skew.

Skewness of the size distributions for *Ophionereis annulata* and *O.esmarki* are different (Figs.8 and 9). The distributions of *O.annulata* tend more towards normal than the more positively skewed distributions of *O.esmarki.* It is reasonable to conclude that *O.esmarki* has better recruitment than *O.annulata,* which is consistent with Hendler's (1975) analysis.

A complication in the comparison of *O.annulata* and *O.esmarki* is that, at False Point, *O.esmarki* is at the southern end of its range whereas *O.annulata* is at its northern end (Ricketts & Calvin 1962). Would the size distributions of *O.esmarki* at Pacific Grove, California, look much like those of *O.annulata* and be more normal and indicate poor recruitment success? Would the size distributions of *O.annulata* in Ecuador be like *O.esmarki* at False Point and show better recruitment success by having size distributions with greater positive skew? Because *O.esmarki* is viviparous, is it free from the forces that modify recruitment in *Strongylocentrotus purpuratus*? Is its recruitment, as a function of latitude, bell-shaped with a maximum somewhere in its midrange as would be predicted by the theoretical considerations of Endler (1977), or is recruitment in viviparous forms a monotonically increasing function from north to south as it seems to be in *Strongylocentrotus* spp?

Size distributions of the deep-sea ophiuroids *Ophiura ljungmani* and *Ophiomusium lymani* show seasonal changes in recruitment (Schoener 1968, Gage & Tyler 1981, 1982). I used Schoener's estimates of average size for two times of year to estimate both growth and mortality parameters, K and Z (Table 5). This particular technique (Ebert 1973) requires assumptions of seasonal recruitment, a seasonally stationary age (size) distribution, Brody-Bertalanffy growth, and a constant survival probability following recruitment.

At about 1000 m (stations #87 and #128 in Table 5), both *Ophiura ljunmani* and *Ophiomusium lymani* have similar survival rates ($Z = 2.7\ yr^{-1}$). This Z-value translates to an annual mortality rate, 1-exp(-Z), equal to 93 %. If the populations are stable and stationary the loss of 93 % of the population is replaced by new recruits. The annual recruitment rate is equal to the annual death rate.

The annual mortality rate decreases with depth for both species to about 79 % per annum for *Ophiura ljunmani* at 2000 m and 61 % for *Ophiumusium lymani.* Gage & Tyler's (1982) data for *O.lymani* at about 2200 m provides a lower survival rate of 34% per annum. For *O.ljungmani* at 2900 m (Gage & Tyler 1981), annual survival is estimated to be 62 %. This decrease in the death rate with depth implies that there also is a decrease in the rate of recruitment with depth. As recruitment of *O.ljungmani* at 2900 m is variable from year to year (Gage & Tyler 1981), it is tempting to speculate that the bathymetric cline of recruitment in the deep sea is similar to the latitudinal cline shown by *Strongylocentrotus purpuratus.* The hypothesis is that recruitment should be greatest at the shallowest portion of the species range, similar to the highest recruitment rates at the more tropical portion of the distribution of *S.purpuratus,* and life should be the shortest at the upper bathymetric limits. A second hypothesis is that recruitment rate should decline with depth and should become less predictable.

The relationship between depth, recruitment and mortality are of considerable interest and could be explored even with species that do not show seasonal recruitment by examining the shapes of size distributions. The hypothesis is that size distributions should become more negatively skewed with depth and approach normal distributions with a mean equal to the average maximum size. Because ophiuroids are such common members of deep-sea communities, they seem to be highly suited to test these hypotheses.

5 RECRUITMENT IN HOLOTHUROIDS

Most population studies of holothuroids have been for species that brood (Rutherford 1973, 1977) or have asexual reproduction (Doty 1977, Ebert 1978, Harriott 1978, Franklin 1980, reviewed by Emson & Wilkie 1980).

Rutherford (1973) examined densities, growth and annual gonad cycle in the brooding holothuroid *Cucumaria pseudocurata* in central California. The relationship between the percent survival during the first month following hatching, and the total biomass is highly significant ($r = -0.91$). Rutherford suggested that competition for space was the mechanism that caused the density-dependent relationship and his data provide the most convincing example in echinoderms of the role of adult density in modifying recruitment.

Asexual reproduction by transverse fission occurs in several species of the genus *Holothuria* (Emson & Wilkie 1980). Doty (1977) concluded that there is no evidence that transverse fission is a significant reproductive process in *H. atra* in Guam based on the fact that no increase in the rate of population growth could be found when there was increased fission. Nor did populations of *H.atra* that he studied increase more rapidly than populations of *H.pervicax,* which does not reproduce asexually. However, the proper question is whether

Table 5. Analysis of size distribution of deep-sea ophiuroids to show annual recruitment rate as a function of depth; assumptions are: constant mortality rate, Brody-Bertalanffy growth and annually stationary size structure; K: the growth rate constant; Z: the mortality coefficient and S_∞ mm: asymptotic size. In a stationary population (r = 0) the annual recruitment rate is equal to the annual mortality rate, 1-exp(-Z).

Organism	Average depth	S_∞ mm	K yr^{-1}	Z yr^{-1}	Annual recruitment rate	Location	Author	Comments
Ophiura ljungmani	1178 m	8.5	0.74	2.67	0.93	Gay Head-Bermuda transect in the western N.Atlantic	Schoener (1968)	Analysis done using technique of Ebert (1973); Station #87, July, is t(1) = 0 with $S_{t(1)}$ = 0.8 m, Station #128, Dec, is t(2) = 0.454 and $S_{t(2)}$ = 3.0 mm; S_R = 0.5 mm
	2100 m	8.5	0.24	1.57	0.79	Gay Head-Bermuda transect in the western N.Atlantic	Schoener (1968)	Station #103, May, is t(1) = 0.454 Atlantic and $S_{t(1)}$ = 1.7 mm; Station #131, Dec., is t(2) = 0.909 and $S_{t(2)}$ = 2.4 mm; S_R = 0.5 mm
	2900 m	9.0	0.15	0.97	0.62	Rockall Trough in the eastern N.Atlantic	Gage & Tyler (1981)	Z-value determined from first four age classes in their Table 5; K = 1n (slope) of their Fig.4
Ophiomusium lymani	1178 m	29.5	0.49	2.68	0.93	Gay Head-Bermuda	Schoener (1968)	Station #87, July, is t(1) = 0 with $S_{t(1)}$ = 1.3 mm; Station #128, Dec., is t(2) = 0.454 with $S_{t(2)}$ = 10.3 mm, S_R = 0.5
	2100 m	29.5	0.29	0.94	0.61	Gay Head-Bermuda	Schoener (1968)	Station #103, May, is t(1) = 0.454 with $S_{t(1)}$ = 7.6 mm; Station #131, Dec, is t(2) = 0.090 with $S_{t(2)}$ = 10.3 mm; S_R = 0.5
	2200 m	37.0	0.34	0.41	0.34	Rockall Trough	Gage & Tyler (1982)	S_∞ and K based on their Table 4. Z is based on their epibenthic sled sample of 28 May 1980 using growth parameters and technique of Ebert (1981).

populations of *H.atra* could persist, that is have r equal to or greater than zero, without transverse fission? The value of r does not have to be positive to make a particular mode of reproduction significant.

The population of *Holothuria atra* at Ananij Island, Enewetak Atoll, appears to be stable as the size structure did not change from 1975 to 1976 (Ebert 1978) and the annual loss rate was estimated to be between 48 and 72%. Although I suggested that annual loss (mortality) rate should be equal to the fission rate, in fact this is not the case in a stationary population.

If N is the fraction of an initial population that does not divide during some time interval and F is the fraction that does divide (i.e. N + F = 1), then, at fission, every dividing individual becomes two individuals and the population changes from 1.0 to N + 2F. The question is, what mortality rate, M, or survival rate, 1-M, will bring the new population back to 1.0?

$$1 = (N + 2F)(1 - M) \quad 1)$$
$$N = 1 - F \quad 2)$$
$$1 = (1 + F)(1 - M) \quad 3)$$

consequently

$$M = F/(1 + F) \quad 4)$$

and

$$F = M/(1 - M) \quad 5)$$

Table 7 summarizes some data on fission from Doty (1977), Harriott (1978) and Ebert (1978). There are two significant features. The fission rate may change during a year (Doty 1977) but not necessarily (Harriott 1978). Recruitment from asexual reproduction in *H.atra* may be continuous or seasonal. Second, the rates of fission can be high and equal or exceed an annual rate of 1.0, as suggested by the data from Heron Island (F = 1.0) and Ananij Island (F = 1.8). These high values mean that on the average, individuals undergo fission once or more each year. The mortality rates also must change and increase with increasing fission – assuming that the populations persist and are not declining or growing.

Doty (1977) suggested that fission seems to be a response to environmental stress; namely, temperature. Fission in response to stress seems to be true for other echinoderms (Emson & Wilkie 1980). If fission is initiated only by heat stress, there probably will be a latitudinal cline of this mode of reproduction and it should be less significant at the higher latitude ends of the geographic ranges of such tropical species as *Holothuria atra*.

6 DISCUSSION

Recruitment in echinoderms shows both temporal and spatial variation. Pat-

Table 6. Asexual reproduction in *Holothuria atra* at Guam, Heron Island (Great Barrier Reef) and Enewetak Atoll. For Guam and Heron Island, annual mortality rate was determined by multiplying six-month survival rates together and subtracting from 1.0; for example for Pago Bay, Guam, annual mortality rate = 1.0 – (0.75)(0.119). Fission rate, F = M/(1.0 – M); L: no evidence of fission; D: in the process of dividing; P: fission products.

Author	Area	Date	Sample size	L	D	P	Total L+D+P/2	F 6 month	M 6 month	Survival 6 month	F 1 yr	M 1 yr
Doty (1977)	Pago Bay, Guam	10 April 1976	238	203	1	34	221	0.081	0.075	0.925		
		10 Dec. 1976	283	172	2	55	229	0.249	0.199	0.801	0.35	0.26
Harriott (1978)	Gutter at Heron Is.	Aug. 1977	51	20		31	36	0.444	0.308	0.692		
		Feb. 1978	52	22		30	37	0.405	0.288	0.712	1.04	0.51
Ebert (1978)	Gutter at Ananij Is.	1975-1976									1.78	0.64

terns include latitudinal clines, as suggested for *Strongylocentrotus purpuratus, S. franciscanus, Acanthaster planci,* and *Asterina gibbosa* and smaller clines as shown for *Asterias forbesi* and *Lytechinus variegatus.* There are clines associated with depth as shown for *Strongylocentrotus purpuratus* in the intertidal, *Asterias forbesi* in the shallow subtidal, and *Ophiura ljungmani* and *Ophiomusium lymani* in the deep sea. Temporal variation is present in virtually all species discussed in this paper including ophiuroids in the deep sea.

Clines in recruitment must have a very significant influence on the evolution of life history traits as different adaptations most certainly must be required at different ends of a long cline that shows unpredictable recruitment at one end and predictable recruitment at the other. Two different life histories would be selected for in the same species in different parts of the geographic range. Clearly, more studies are needed that focus on clinal variation in recruitment as well as other life-history features.

Most parts of the life cycle are important in influencing recruitment success. Adult density has a negative effect on recruitment of *Cucumaria pseudocurata* (Rutherford 1973). Timko (1979) argued that the interaction is negative for the sand dollar *Dendraster excentricus* but her results could not be repeated by Chia & Atwood (1982). Birkeland & Mesmer (1978) and Highsmith (1977) presented data that indicate the relationship is positive for the same species. Tegner & Dayton (1977) argued that there is a significant positive influence of density of *Strongylocentrotus franciscanus* on recruitment. In all of these cases, the importance of adult density is at settlement or immediately following settlement. This is in contrast with the arguments that breeding density promotes recruitment in *Acanthaster planci* (e.g. Vine 1973). Although reasonable, there is no evidence to show that aggregation affects the number of eggs that are fertilized or that this, in turn, influences settlement or recruitment. There also is no evidence that stock size of echinoderms influences recruitment other than the trivial case where a stock of zero size produces zero recruits. The relationship shown between settlement and adult density for *Asterias forbesi* shows no obvious relationship.

Events in the plankton are difficult to study and there are major technical problems in evaluating the importance of predators in determining settlement density of echinoderms. After Brooks & Dodson (1965), it is difficult to think of the plankton in terms other than fairly tight interactions; however, a test of the effectiveness of small marine predators in limiting their prey has yet to be experimentally demonstrated in the field. Freshwater biologists have had considerable success with adjusting zooplankton composition in large field enclosures (e.g. Neill & Peacock 1980, Kerfoot & DeMott 1980) but problems associated with the construction of enclosures that are scaled-up to approximate natural conditions in marine environments have had mixed results (e.g. McAllister *et al.* 1961, Grice *et al.* 1980) and suggest that technical difficulties are formidable in attempting to test the role of planktonic predators in influencing recruitment.

Early survival following settlement is poorly understood but it seems reasonable that tiny predators must consume many small echinoderms. Techniques for culturing echinoderm larvae to the point of metamorphosis are well developed (Hinegardner 1967, Lucas 1974, Yamaguchi 1974b) and it may be possible to devise direct tests of the influence of small predators. A simple test of releasing newly metamorphosed larvae in clean vs. aged rubble would be a first test of the role of small predators. If survivorship is poorer in aged rubble then, in principle, it should be possible to manipulate the small predators that live in the rubble to determine which are the most important.

The influence of water movement in determining recruitment is clear for *Strongylocentrotus purpuratus* though the exact mechanism is not. The influence of water movement may never be directly testable but a better understanding of larval autecology would permit certain indirect tests such as: Along a coastline, species with similar precompetent and competent periods will have coincident years of good and poor recruitment if physical transport is significant in causing year to year variation. Clearly, more must be known about larval biology to make such tests.

Echinoderms are abundant in all seas and at all depths. As a consequence they provide readily available material for study and are a reasonable group of organisms to use to answer basic questions about recruitment in marine communities and about the evolution of life histories along clines. Documentation of patterns and an approach to explaining the patterns that emphasizes hypothesis testing will continue to provide stimulating research for echinoderm biologists.

7 ACKNOWLEDGEMENTS

Portions of this work were supported by NSF OEC75-10442 to study life history evolution. Some of the echinoid data at Sunset Bay, Oregon, were gathered while I was supported by NSF GB-977 to Peter Frank. The analysis of covariance program is part of the BMDP programs developed at the Health Sciences Computing Facility, UCLA, and sponsored by NIH Special Resources Grant RR-3. The manuscript benefitted from discussions in my graduate seminar in aquatic ecology (Fall 1981) and the comments of C.Barilotti, C.Birkeland and D.Dexter and particularly from the critical evaluation of J.Lawrence.

8 REFERENCES

Achituv, Y. 1969. Studies on the reproduction and distribution of *Asterina burtoni* Gray and *A.wega* Perrier (Asteroidea) in the Red Sea and the eastern Mediterranean. *Israel J. Zool.* 18:329-342.

Bakun, A. 1973. Coastal upwelling indices, West coast of North America, 1946-71. *NOAA Tech. Rpt. NMFS*, SSRF671:103pp.

Barker, M.F. 1977. Observations on the settlement of the brachiolaria larvae of *Stichaster australis* (Verrill) and *Coscinasterias calamaria* (Gray) (Echinodermata: Asteroidea) in the laboratory and on the shore. *J. exp. mar. Biol. Ecol.* 30:95-108.

Bernard, F.R. & D.C.Miller 1973. Preliminary investigation of the red sea urchin of British Columbia, *Strongylocentrotus franciscanus* (Agassiz). *Tech. Rpts Fish. Res. Bd. Canada* 400:37pp.

Bernstein, B.B. & N.Jung 1979. Selective pressures and coevolution in a kelp canopy community in southern California. *Ecol. Monog.* 49:335-355.

Birkeland, C. 1974. Interactions between a seapen and seven of its predators. *Ecol. Monog.* 44:211-232.

Birkeland, C. 1979. Report on the *Acanthaster planci* (Rrusech) survey of Palau 18 to 26 May 1979. *Rpt to Mar. Res. Div. Trust Territory of the Pacific Islands* (U.Guam Marine Laboratory), 30pp.

Birkeland, C. 1982. Terrestrial runoff as a cause of outbreaks of *Acanthaster planci* (Echinodermata: Asteroidea). *Mar. Biol.* 69:175-185.

Birkeland, C., F.-S.Chia & R.R.Strathmann 1971. Development, substratum selection, delay of metamorphosis and growth in the seastar, *Mediaster aequalis* Stimpson. *Biol. Bull.* 141:99-108.

Birkeland, C. & K.E.Mesmer 1978. An experiment on the role of historical chance in the distribution of populations of *Dendraster excentricus*. International Conference on Echinoderm Biology, Sydney, NSW, Australia. March 12-17, 1978 (Abstract).

Branham, J.M., S.A.Reed, J.H.Bailey & J.Caperon 1971. Coral-eating sea stars *Acanthaster planci* in Hawaii. *Science* 1972:1155-1157.

Brooks, J.L. & S.I.Dodson 1965. Predation, body size, and composition of plankton. *Science* 150:28-35.

Buchanan, J.M. 1967. Dispersion and demography of some infaunal echinoderm populations. *Symp. zool. Soc. Lond.* 20:1-11.

Burkenroad, M.D. 1957. Intensity of setting of starfish in Long Island Sound in relationship to fluctuations of the stock of adult starfish and in the setting of oysters. *Ecology* 38:164-165.

Cairns, J.L. 1968. Thermocline strength fluctuations in coastal waters. *J. Geophy. Res.* 73:2591-2595.

Cameron, A.M. & R.Endean 1981. Renewed population outbreaks of a rare and specialized carnivore (the starfish *Acanthaster planci*) in a complex high-diversity system (the Great Barrier Reef). *Proc. 4th International Conf. Coral Reefs* (in press).

Cameron, R.A. & S.C.Schroeter 1980. Sea urchin recruitment: Effect of substrate selection on juvenile distribution. *Mar. Ecol. Prog. Ser.* 2:243-247.

Camp, D.K., S.P.Cobb & J.F.van Breedvelt 1973. Overgrazing of seagrasses by a regular urchin, *Lytechinus variegatus. Biosci.* 23:37-38.

Campbell, A.C. & R.F.G.Ormond 1970. The threat of the 'Crown-of-thorns' starfish *(Acanthaster planci)* to coral reefs in the Indo-Pacific area: Observations on a normal population in the Red Sea. *Biol. Conserv.* 2:246-251.

Chelton, D.B. 1981. Interannual variability of the California current – physical factors. *CALCOFI Rep.* 22:34-48.

Chesher, R.H. 1969. Destruction of Pacific corals by the sea star *Acanthaster planci. Science* 165:280-283.

Chia, F.-S. 1966. Brooding behavior of a six-rayed starfish, *Leptasterias hexactis. Biol. Bull.* 130:304-325.

Chia, F.-S. 1978. Viviparity in echinoderms. International Conference on Echinoderm Biology, Sydney, NSW, Australia, March 12-17, 1978. (Abstract).

Chia, F.-S. & D.G.Atwood 1982. Pigment cells in the jelly coat of sand dollar eggs. In: J.M.Lawrence (ed.), *Echinoderms: Proc. intern. Conf., Tampa Bay.* pp.481-484. Balkema, Rotterdam.

Coe, W.R. 1956. Fluctuations in populations of littoral marine invertebrates. *J. mar. Res.* 15:212-232.

Crump, R.G. & R.H.Emson 1978. Some aspects of the population dynamics of *Asterina gibbosa* (Asteroidea). *J. mar. biol. Ass. UK* 58:451-466.

Cushing, D.H. 1971. Upwelling and the production of fish. *Adv. mar. Biol.* 9:255-334.

Cushing, D.H. & J.G.K.Harris 1973. Stock and recruitment and the problem of density-independence. *Rapp. P-v. Réun. Cons. perm. int. Explor. Mer.* 164:142-155.

Dana, T.F., W.A.Newman & E.W.Fager 1972. *Acanthaster* aggregations: interpreted as primarily responses to natural phenomena. *Pac. Sci.* 26:355-372.

Dartnall, A.J. 1969. A viviparous species of *Patiriella* (Asteroidea, Asterinidae) from Tasmania. *Proc. Linn. Soc. NSW* 93:294-296.

Dehnel, P.A. 1955. Rates of growth of gastropods as a function of latitude. *Physiol. Zoöl,* 28:115-144.

Dexter, D.M. 1977. A natural history of the sand dollar *Encope stokesi* L. Agassiz in Panama. *Bull. mar. Sci.* 27:544-551.

Dixon, W.J. & M.B.Brown (eds) 1979. *BMDP-79. Biomedical computer Programs P-series.* University of California Press, Berkeley. 880pp.

Doty, J.E. 1977. Fission in *Holothuria atra* and holothurian population growth. M.Sc. Thesis, University of Guam, Agana. 54pp.

Duggins, D.O. 1980. Kelp dominated communities: experimental studies on the relationships between sea urchins, their predators and their algal resources. Ph.D. Dissert., University of Washington, Seattle.

Ebert, T.A. 1968. Growth rates of the sea urchin *Strongylocentrotus purpuratus* related to food availability and spine abrasion. *Ecology* 49:1075-1091.

Ebert, T.A. 1973. Estimating growth and mortality rates from size data. *Oecologia* 11: 281-298.

Ebert, T.A. 1978. Growth and size of the tropical sea cucumber *Holothuria (Halodeima) atra* Jäger at Enewetak Atoll, Marshall Islands. *Pac. Sci.* 32:183-190.

Ebert, T.A. 1981. Estimating mortality from growth parameters and a size distribution when recruitment is periodic. *Limnol. Oceanogr.* 26:764-769.

Ebert, T.A. & D.M.Dexter 1975. A natural history study of *Encope grandis* and *Mellita grantii,* two sand dollars in the northern Gulf of California, Mexico. *Mar. Biol.* 32:397-407.

Ebling, F.J., A.D.Hawkins, J.A.Kitching, L.Muntz & V.M.Pratt 1966. The ecology of Lough Ine XVI. Predation and diurnal migration in the *Paracentrotus* community. *J. anim. Ecol.* 35:559-566.

Emson, R.H. & R.G.Crump 1979. Description of a new species of *Asterina* (Asteroidea), with an account of its ecology. *J. mar. biol. Ass. UK* 59:77-94.

Emson, R.H. & I.C.Wilkie 1980. Fission and autotomy in echinoderms. *Oceanogr. mar. Biol. ann. Rev.* 18:155-250.

Endean, R. 1973. Population explosions of *Acanthaster planci* and associated destruction of hermatypic corals in the Indo-west Pacific region. In: O.A.Jones & R.Endean (eds), *Biology and Geology of Coral Reefs,* vol.II, Biology 1, pp.389-438. Academic Press, New York.

Endean, R. 1974. *Acanthaster planci* on the Great Barrier Reef. *Proc. 2nd Intl Symp. Coral Reefs,* pp.563-576.

Endean, R. 1976. Destruction and recovery of coral reef communities. In: O.A.Jones & R.Endean (eds), *Biology and Geology of Coral Reefs,* vol.III, Biology 2, pp.215-254. Academic Press, New York.

Endean, R. & R.H.Chesher 1973. Temporal and spatial distribution of *Acanthaster planci* population explosions in the Indo-west Pacific region. *Biol. Conserv.* 5:87-95.

Endean, R. & W.Stablum 1973. A study of some aspects of the crown-of-thorns starfish *(Acanthaster planci)* infestations of reef of Australia's Great Barrier Reef. *Atoll Res. Bull.* 167:1-76.

Endler, J.A. 1977. *Geographic variation, speciation, and clines.* Monog. Pop. Biol. 10. Princeton Univ. Press, Princeton. 246pp.

Fariás Sánchez, J.A. 1980. Observaciones preliminares sobre una poblacion de los erizos comerciales *Strongylocentrotus franciscanus* (Echinodermata, Echinozoa) en Baja California, Mexico. Thesis, Escuela Superior de Ciencias Marinas, Univ. Autonoma de Baja California, Ensenada. 94pp.

Farmanfarmaian, A. & A.C.Giese 1963. Thermal tolerance and acclimation in the western purple sea urchin, *Strongylocentrotus purpuratus. Physiol. Zoöl.* 36:237-243.

Ford, R.F., D.G.Foreman, K.J.Grubbs, D.B.Innis, D.C.Sommerville & F.L.Steinert 1976. *Effects of thermal effluent on marine organisms.* Tech. Rpt, Center for Marine Studies, San Diego State University, San Diego. 310pp.

Foreman, R.E. 1977. Benthic community modification and recovery following intensive grazing by *Strongylocentrotus droebachiensis. Helgolander wiss. Meeresunters.* 30: 468-484.

Frank, P.W. 1975. Latitudinal variation in the life history features of the black turban snail *Tegula funebralis* (Prosobranchia: Trochidae). *Mar. Biol.* 31:181-192.

Frankel, E. 1975. *Acanthaster* in the past: evidence from the Great Barrier Reef. *Proc. crown-of-thorns Starfish Seminar,* pp.159-165. Aust. Govt Publ. Service, Canberra.

Frankel, E. 1977a. Evidence from the Great Barrier Reef of ancient *Acanthaster* aggregations. *Atoll Res. Bull.* 220:75-93.

Frankel, E. 1977b. Previous *Acanthaster* aggregations in the Great Barrier Reef. *Proc. 3rd Internat. Coral Reef Symp.* pp.201-208.

Franklin, S.E. 1980. The reproductive biology and some aspects of the population ecology of the holothurians *Holothuria leucospilota* (Brandt) and *Stichopus chloronotus* (Brandt). Ph.D. Thesis, University of Sydney, Australia.

Fuji, A. & K.Kawamura 1970. Studies on the biology of the sea urchin VI. Habitat structure and regional distribution of *Strongylocentrotus intermedius* on a rocky shore of southern Hokkaido. *Bull. Jap. Soc. scient. Fish.* 36:755-762.

Gage, J.D. & P.A.Tyler 1981. Re-appraisal of age composition, growth and survivorship of the deep-sea brittle star *Ophiura ljungmani* from size structure in a sample time series from the Rockall Trough. *Mar. Biol.* 64:163-172.

Gage, J.D. & P.A.Tyler 1982. Growth and reproduction of the deep-sea brittlestar *Ophiomusium lymani* Wyville Thomson. *Oceanol. Acta.* 5.

Galtsoff, P.S. & V.L.Loosanoff 1939. Natural history and method of controlling the starfish (*Asterias forbesi,* Desor). *Bull. Bur. Fish., Wash.* 49:75-132.

Gemmill, J.F. 1914. The development and certain points in the adult structure of the starfish *Asterias rubens* L. *Phil. Trans. Roy. Soc.* (B)205:213-294.

Gould, S.J. & R.C.Lewontin 1979. The spandrels of San Marco and the Panglossian paradigm: a critique of the adaptationist programme. *Proc. Roy. Soc. London* (B)205:581-598.

Grice, G.D., R.P.Harris, M.R.Reeve, J.F.Heinbokel & C.O.Davis 1980. Large-scale enclosed water-column ecosystem. An overview of Foodweb I, the final CEPEX experiment. *J. mar. biol. Ass. UK* 60:401-414.

Harriott, V. 1978. Sexual and asexual reproduction of *Holothuria atra* (Jaeger) at Heron Island reef, Great Barrier Reef. Manuscript, 13pp. (University of Queensland, Brisbane).

Hayman, R.A. & A.V.Tyler 1980. Environment and cohort strength of Dover sole and English sole. *Trans. Am. Fish. Soc.* 109:54-70.

Hendler, G. 1975. Adaptational significance of the patterns of ophiuroid development. *Amer. Zool.* 15:691-715.

Hendler, G. 1982. An echinoderm vitellaria with a bilateral larval skeleton: evidence for the evolution of ophiuroid vitellariae from ophioplutei. *Biol. Bull.* 163:431-437.

Highsmith, R.C. 1977. Larval substrate selection, metamorphosis, and mortality in the sand dollar, *Dendraster excentricus. Am. Zool.* 17:935.

Himmelman, J.H. 1980. Synchronization of spawning in marine invertebrates by phytoplankton. In: W.H.Clark, jr & T.S.Adams (eds), *Advances in invertebrate Reproduction*, pp.3-19. Elsevier North-Holland Inc, New York.

Hinegardner, R.T. 1967. Echinoderms. In: F.H.Wilt & N.K.Wessels (eds), *Methods in Developmental Biology*, pp.139-155. Thomas Y.Crowell Co, New York.

Hinegardner, R.T. & M.M.R.Tuzzi 1981. Laboratory culture of the sea urchin *Lytechinus pictus.* In: National Research Council Committee on Marine Invertebrates (eds), *Laboratory Animal Management. Marine Invertebrates*, pp.291-302. Nat. Acad. Press, Washington, DC. 382pp.

Hjort, J. 1914. Fluctuations in the great fisheries of northern Europe, viewed in the light of biological research. *Rapp. P-v. Réun. Cons. perm. int. Explor. Mer* 10:1-228.

Hobson, E.S. & J.R.Chess 1976. Trophic interactions among fishes and zooplankters near shore at Santa Catalina Island, California. *Fish. Bull.* 74:567-598.

Iles, T.D. 1973. Interaction of environment and parent stock size in determining recruitment in the Pacific sardine as revealed by analysis of density-dependent O-group growth. *Rapp. P-v. Réun. Cons. perm. int. Explor. Mer* 164:228-240.

Jackson, G.A. & R.R.Strathmann 1981. Larval mortality from offshore mixing as a link between precompetent and competent periods of development. *Amer. Nat.* 118:16-26.

Kamykowski, D.L. 1972. Some physical and chemical aspects of the phytoplankton ecology of La Jolla Bay. Ph.D. Dissert, Univ. California, San Diego. 269pp.

Kenchington, R.A. 1977. Growth and recruitment of *Acanthaster planci* (L.) on the Great Barrier Reef. *Biol. Conserv.* 11:103-118.

Kerfoot, W.C. & W.R.DeMott 1980. Foundations for evaluating community interactions: The use of enclosures to investigate coexistence of *Daphnia* and *Bosmina.* In: W.C.Kerfoot (ed.), *Evolution and Ecology of Zooplankton Communities*, Species Symp. vol.3, pp.725-741. Amer. Soc. Limnol. Oceanog., Univ. Press of New England, Hanover.

Kinne, O. 1970. Temperature. Animals. Invertebrates. In: O.Kinne (ed.), *Marine Ecology*, vol.1(1), pp.407-514. Wiley-Interscience, New York.

Lane, J.E.M. & J.M.Lawrence 1980. Seasonal variation in body growth, density and distribution of a population of sand dollars, *Mellita quinquiesperforata* (Leske). *Bull. mar. Sci.* 30:871-882.

Lang, C. & K.H.Mann 1976. Changes in sea urchin populations after the destruction of kelp beds. *Mar. Biol.* 36:321-326.

Lawson-Kerr, C. & D.T.Anderson 1978. Reproduction, spawning and development of the starfish *Patiriella exigua* (Lamarck) (Asteroidea: Asterinidae) and some comparisons with *P.calcar* (Lamarck). *Aust. J. mar. freshwater Res.* 29:45-53.

Laxton, J.H. 1974. Aspects of the ecology of the coral-eating starfish *Acanthaster planci. Biol. J. Linn. Soc.* 6:19-45.

Loosanoff, V.L. 1964. Variation in time and intensity of setting of the starfish, *Asterias forbesi*, in Long Island Sound during a twenty-five year period. *Biol. Bull.* 126:423-439.

Low, C.J. 1975. The effect of grouping of *Strongylocentrotus franciscanus*, the giant red

sea urchin, on its population biology. Ph.D. Dissert. Univ. British Columbia. 157pp.
Lucas, J.S. 1973. Reproductive and larval biology of *Acanthaster planci* in Great Barrier Reef waters. *Micronesica* 9:197-207.
Lucas, J.S. 1974. Environmental influences of the early development of *Acanthaster planci* (L.). *Proc. Crown-of-thorns Starfish Seminar,* pp.109 -121. Aust. Govt Publ. Service, Canberra.
MacArthur, R.H. & E.O.Wilson 1967. *The Theory of Island Biogeography.* Princeton Univ. Press, Princeton. 203pp.
MacDonald, P.D.M. & T.J.Pitcher 1979. Age-groups from size-frequency data: A versatile and efficient method of analyzing distribution mixtures. *J. Fish. Res. Bd. Canada,* 36:987-1001.
Matsui, I. 1966. *The Propagation of Sea Urchins.* Booklet No.12, Nippon Suisan Shigan Hogo Kyokai (Japan Fisheries Research Conservation Association). Fisheries Board of Canada Transl. Ser. No.1063, 105pp.
Mauzey, K.P., C.Birkeland & P.K.Dayton 1968. Feeding behavior of asteroids and escape responses of their prey in the Puget Sound region. *Ecology* 49:603-619.
McAllister, C.D., T.R.Parsons, K.Stephens & J.D.H.Strickland 1961. Measurements of primary production in coastal sea water using a large-volume plastic sphere. *Limnol. Oceanog.* 6:237-258.
Menge, B.A. 1975. Brood or broadcast? The adaptive significance of different strategies in the two intertidal sea stars *Leptasterias hexactis* and *Pisaster ochraceus. Mar. Biol.* 31:87-100.
Menge, J.L. & B.A.Menge 1974. Role of resource allocation, aggression and spatial heterogeneity in coexistence of two competing intertidal starfish. *Ecol. Monog.* 44:189-209.
Moore, H.B., T.Jutare, J.C.Bauer & J.A.Jones 1963. The biology of *Lytechinus variegatus. Bull. mar. Sci.* 13:23-53.
Murphy, G.I. 1968. Patterns in life history and the environment. *Amer. Nat.* 102:391-411.
Muscat, A.M. 1975. Reproduction and growth in the ophiuroid, *Ophionereis annulata.* M.Sc. Thesis, San Diego State University, San Diego. 85pp.
Neill, W.E. & A.Peacock 1980. Breaking the bottleneck: Interaction of invertebrate predators and nutrients in oligotrophic lakes. In: W.C.Kerfoot (ed.), *Evolution and ecology of Zooplankton Communities.* Special Symp. vol.3, pp.715-724. Am. Soc. Limnol. Oceanog., Univ. Press of New England, Hanover.
North, W.J. 1974. A review of the studies supporting sea urchin control as a means of restoring kelp beds. In: *Kelp Habitat Improvement Project,* Annual Rept. (1 July 1973-30 June 1974), Calif. Institute Tech. Pasadena. pp.95-108.
Oguro, C., M.Komatsu & Y.T.Kano 1978. Development of a viviparous sea-star *Patiriella pseudoexigua pacifica* Hayashi. International Conference on Echinoderm Biology, Sydney, NSW, Australia. March 12-17, 1978 (Abstract).
Ormond, R.F.G. & A.C.Campbell 1974. Formation and breakdown of *Acanthaster planci* aggregations in the Red Sea. *Proc. 2nd intern. Symp. Coral Reefs.* pp.595-619.
Pearse, J.S., M.E.Clark, D.L.Leighton, C.T.Mitchell & W.J.North 1970. Final report. Marine waste disposal and sea urchin ecology. Appendix. In: *Kelp Habitat Improvement Project,* annual report (1 July 1969-30 June 1970) pp.1-93. Calif. Institute of Technology, Pasadena.
Pearse, J.S. & V.B.Pearse 1975. Growth zones in echinoid skeleton. *Amer. Zool.* 15:731-753.
Pearson, R.G. & R.Endean 1969. A preliminary study of the coral predator *Acanthaster planci* (L.) (Asteroidea) on the Great Barrier Reef. *Fish. Notes, Brisbane,* 3:27-55.
Potts, D.C. (in press). Crown-of-thorns starfish – man-induced pest or natural phenomenon? In: R.L.Kitching & R.E.Jones (eds), *The Ecology of Pests,* pp.55-86. CSIRO Publ. Serv., Canberra.

Randall, J.E. 1972. Chemical pollution in the sea and the crown-of-thorns starfish *(Acanthaster planci). Biotropica* 4:132-144.
Reaka, M. L. 1979. The evolutionary ecology of life history patterns in stomatopod Crustacea. In: S.E.Stancyk (ed.), *Reproductive Ecology of marine Invertebrates.* pp.235-260. Belle W.Baruch Library. Univ. South Carolina Press, Columbia.
Ricketts, E.F. & J.Calvin 1962. *Between Pacific Tides.* 3rd edition revised by J.W.Hedgpeth. Stanford Univ. Press, Stanford. 516pp.
Rivera, J.A. 1979. Aspects of the biology of *Lytechinus variegatus* (Lamarck, 1816) at Jobos Bay, Puerto Rico (Echinoidea: Toxopneustiae). M.Sc. Thesis, Univ. Puerto Rico, Mayaguez. 192pp.
Roff, D.A. 1981. Reproductive uncertainty and the evolution of iteroparity: Why don't flatfish put all their eggs in one basket? *Can. J. Fish. aquat. Sci.* 38:968-977.
Rutherford, J.C. 1973. Reproduction, growth and mortality of the holothurian *Cucumaria pseudocurata. Mar. Biol.* 22:167-176.
Rutherford, J.C. 1977. Variation in egg numbers between populations and between years in the holothurian *Cucumaria curata. Mar. Biol.* 43:175-180.
Sale, P.F., D.C. Potts & E.Frankel 1976. Recent studies on *Acanthaster planci.* A report to the council of the Great Barrier Reef Committee. *Search* 7:334-338.
Schoener, A. 1968. Evidence for reproductive periodicity in the deep sea. *Ecology* 49: 81-87.
Schroeter, S.C. 1978. Experimental studies of competition as a factor affecting the distribution and abundance of purple sea urchins, *Strongylocentrotus purpuratus* (Stimpson). Ph.D. Dissert., Univ. California, Santa Barbara. 184pp.
Sloan, N.A. 1980. Aspects of the feeding biology of asteroids. *Oceanogr. mar. Biol. ann. Rev.* 18:57-124.
Slobodkin, L.B. & A.Rapoport 1974. An optimal strategy of evolution. *Quart. Rev. Biol.* 49:181-200.
Smith, R.L. 1968. Upwelling. *Oceanogr. mar. Biol. ann. Rev.* 6:11-46.
Stearns, S.C. 1976. Life history tactics: a review of the ideas. *Quart. Rev. Biol.* 51:3-47.
Stearns, S.C. 1977. The evolution of life history traits. *Ann. Rev. Ecol. Systematics* 8: 145-171.
Strathmann, R. 1978. Length of pelagic period in echinoderms with feeding larvae from the northeast Pacific. *J. exp. mar. Biol. Ecol.* 34:23-27.
Strickland, J.D. (ed.) 1970. The ecology of plankton off La Jolla, California, in the period April through September 1966. *Bull. Scripps Instn. Oceanogr.* 17.
Sutcliffe, W.H. jr 1973. Correlations between seasonal river discharge and local landings of American lobster *(Homarus americanus)* and Atlantic halibut *(Hippoglossus hippoglossus)* in the Gulf of St Lawrence. *J. Fish. Res. Bd. Canada* 30:856-859.
Tegner, M.J. & P.K.Dayton 1977. Sea urchin recruitment patterns and implications of commercial fishing. *Science* 196:324-326.
Tegner, M.J. & P.K.Dayton 1981. Population structure, recruitment and mortality of two sea urchins (*Strongylocentrotus franciscanus* and *S.purpuratus*) in a kelp forest. *Mar. Ecol. Prog. Ser.* 5:255-268.
Thorson, G. 1950. Reproductive and larval ecology of marine bottom invertebrates. *Biol. Rev.* 25:1-45.
Timko, P. 1979. Larviphagy and oophagy in benthic invertebrates: a demonstration for *Dendraster excentricus* (Echinoidea). In: S.E.Stancyk (ed.), *Reproductive ecology of marine invertebrates,* pp.91-98. Belle W.Baruch Library. Univ. South Carolina Press, Columbia.
Tyler, A. 1936. On the energetics of differentiation IV. Comparison of the rates of oxygen consumption and of development at different temperatures of eggs of some marine animals. *Biol. Bull.* 71:82-100.

Vine, P.J. 1973. Crown of thorns *(Acanthaster planci)* plagues: The natural causes theory. *Atoll. Res. Bull.* 166:1-14.

Weber, J.N. & P.M.J.Woodhead 1970. Ecological studies of the coral predator *Acanthaster planci* in the south Pacific. *Mar. Biol.* 6:12-17.

Yamaguchi, M. 1973a. Early life histories of coral reef asteroids, with special reference to *Acanthaster planci* (L.) In: O.A.Jones & R.Endean (eds), *Biology and Geology of Coral Reefs.* Vol.II, Biology 1, pp.369-387. Academic Press, New York.

Yamaguchi, M. 1973b. Recruitment of coral reef asteroids, with emphasis on *Acanthaster planci* (L.). *Micronesica* 9:207-212.

Yamaguchi, M. 1974. Growth of juvenile *Acanthaster planci* in the laboratory. *Pacific Sci.* 28:123-138.

Yamaguchi, M. 1975a. Coral-reef asteroids of Guam. *Biotropica* 7:12-23.

Yamaguchi, M. 1975b. Estimating growth parameters from growth rate data. Problems with marine sedentary invertebrates. *Oecologia* 20:321-332.